Discovering Higher Mathematics

Discovering Higher Mathematics

Four Habits of Highly Effective Mathematicians

Alan Levine

Franklin and Marshall College

HARCOURT
ACADEMIC
PRESS

San Diego   San Francisco   New York   Boston
London   Toronto   Sydney   Tokyo

ACADEMIC PRESS
A Harcourt Science and Technology Company
525 B Street, Suite 1900, San Diego, CA 92101-4495, USA
http://www.academicpress.com

Academic Press
24–28 Oval Road, London NW1 7DX United Kingdom
http://www.hbuk.co.uk/ap/

Harcourt/Academic Press
200 Wheeler Road, Burlington, MA 01803, USA
http://www.harcourt-ap.com

Library of Congress Catalog Number: 99-65103

ISBN: 0-12-445460-7

Printed in the United States of America
99 00 01 02 03   QW   6 5 4 3 2 1

In honor of my father, Arthur, and in memory of my mother, Arlene. They taught me algebra, Fibonacci numbers, and other neat mathematics when I was in elementary school, thus starting me on my journey toward "discovering higher mathematics."

Contents

Preface   ix
Acknowledgments   xiii
Tips on Writing Mathematics   xv

1. Preliminary Ideas

1.1.   Properties of the Integers   *1*
1.2.   Conjectures and Proofs About Sequences   *9*
1.3.   Mathematical Induction   *15*
1.4.   Primes and the Fundamental Theorem of Arithmetic   *17*
1.5.   Additional Questions   *25*

2. Numbers and Numerals

2.1.   Rational Numbers   *29*
2.2.   Irrational Numbers   *35*
2.3.   Continued Fractions   *37*
2.4.   Systems of Numeration   *43*
2.5.   Additional Questions   *47*

3. Polynomials and Complex Numbers

3.1.   Polynomials   *51*
3.2.   Complex Numbers   *58*
3.3.   Roots of Polynomials   *61*
3.4.   Geometric Constructions   *67*
3.5.   Additional Questions   *72*

4. Combinatorics and Graph Theory

4.1	Combinations and the Binomial Theorem	75
4.2.	Generating Functions	84
4.3.	Introduction to Graph Theory	87
4.4.	Additional Questions	93

5. Difference Equations and Iteration

5.1.	Linear Difference Equations	97
5.2.	Linear Function Iteration	101
5.3.	Nonlinear Function Iteration	106
5.4.	Additional Questions	111

6. Additional Topics

6.1.	Arithmetic and Geometric Means	115
6.2.	Greatest Integer Function	119
6.3.	Sums and Differences	122
6.4.	Diophantine Equations and the Chinese Remainder Theorem	125
6.5.	Pythagorean Triplets	130
6.6.	Fermat's Little Theorem	134
6.7.	Additional Questions	136

7. Topics in Algebra

7.1.	Rings	141
7.2.	Unique Factorization	147
7.3.	Fields	155
7.4.	Additional Questions	158

Appendix: An Introduction to Symbolic Logic

A.1.	Propositions	161
A.2.	Open Sentences and Quantifiers	166
A.3.	Additional Questions	169

Index

171

Preface

One of the problems with traditional mathematics education is that students often leave their courses with a distorted view of what mathematics really is. Typically, students learn mathematics passively, by listening to an instructor lecture and present examples. Then, they practice similar examples for homework and are tested by having to do more of the same. Not only does this lead students to believe that mathematics can be learned simply by repetition, it also makes mathematics appear to be a boring, lifeless subject.

Students also have little idea of how mathematics was created. Textbooks contain the result of years of work by the authors, with all the mistakes and false starts removed. This gives the impression that mathematics was created by divine inspiration. Or, perhaps, mathematics just appeared on Earth one day (sort of a "big bang" theory of mathematics), where it was discovered by some old Greeks who started naming things after themselves (the Pythagorean theorem, Euclidean geometry, etc.).

> *Far too often, mathematics in the classroom is a freeze-dried mathematics—rigid, cold, and unappealing. Instead of exploration, there is drill; instead of investigation, imitation. From elementary school arithmetic to college calculus, mathematics in the classroom is dramatically different from mathematics in practice.*[1]

In this text, we will try to change these impressions by presenting mathematics as a living, breathing subject that is constantly being developed and refined. The emphasis will not be on learning a specific mathematical content (such as calculus or algebra) but rather on *creating* mathematics.

Creating mathematics consists of four main processes. *Experimentation* involves generating data either by hand calculation or with a calculator or computer

[1] Steen, L. A., "Celebrating Mathematics," *American Mathematical Monthly,* vol. 95, no. 5 (1988).

program. The experiment may involve looking at special cases or at simpler, related problems. *Conjecture* is the process of looking at the results of the experiment, combining them with some intuition, and proposing a hypothesis. This is often the most difficult part of the procedure. *Proof* is the process of verifying that the conjecture is indeed true. Conversely, failure to construct a proof might show why the conjecture is false. Finally, *generalization* involves taking the results just proven and extending them to other problems. Nearly every new result in mathematics is a generalization or adaptation of other, previously proven results.

Of course, it is not possible to learn these processes in the absence of content and, consequently, we will cover a variety of mathematical topics such as number theory, polynomials, combinatorics, and discrete mathematics. These topics were chosen because they involve such concepts as integers, functions, sets, and arithmetic, which are already familiar to most students at this level, and because they readily lend themselves to each of the processes described above. Although we will look at several interesting problems under each topic, we will not do a complete course in any of these areas.

To see how these processes work in practice, consider the following problem: In how many ways can a positive integer n be written as the sum of two other positive integers, where the order in which the integers are added matters?

To solve this problem, we begin by experimenting—that is, by trying out some simple cases. If $n = 3$, there are two ways ($1 + 2$ and $2 + 1$). If $n = 4$, there are three ways ($1 + 3$, $2 + 2$, and $3 + 1$). If $n = 5$, there are four ways ($1 + 4$, $2 + 3$, $3 + 2$, and $4 + 1$). Based on this evidence, it would appear that there are $n - 1$ ways of expressing n as the sum of two positive integers. That's the conjecture. To prove the conjecture, we note that the first integer in the sum can be any of the integers $\{1, 2, 3, \ldots, n - 1\}$. Since the choice of the first integer uniquely determines the second integer, there really are $n - 1$ different sums.

There are several generalizations. For instance, what happens if the order doesn't matter—that is, if $2 + 3$ and $3 + 2$ are considered the same? What happens if we express n as the sum of three integers? Or as the sum of even integers? Some of the generalizations may be relatively straightforward; others may be deceptively difficult.

To master these processes, students must change their approach to learning. They cannot merely sit back and wait for the instructor to tell them what experiments to do and what the correct conjectures, proofs, and generalizations are. They must *actively participate* in the learning process.

This text is structured to encourage active student participation. Unlike most textbooks, there are no complete discourses on the subject matter. Consequently, the student cannot learn the material just by reading the book or by listening to an instructor lecture. Rather, each chapter in the book contains a series of individual modules. The modules consist of some introductory material, followed by a number of questions interwoven with explanations and other relevant information. The students must answer the questions to "fill in the missing pieces." To ensure that students participate conscientiously in this process and derive maximum benefit, there is no answer key. (However, there is a teacher's edition containing the answers to all the questions.)

Finally, since one of the main themes of the text is proof, there is a considerable amount of writing for students to do. Not only are there the questions within the text to answer, there are also additional questions at the end of each chapter that can be used as homework. At this point, students have not had much experience writing mathematics; hence, it is appropriate to spend time discussing stylistic issues. A guide covering some of these issues is included in the text.

Acknowledgments

I would like to thank the many people who have contributed significantly to the development of this text.

In the spring of 1995, my colleagues in the Mathematics Department at Franklin and Marshall College—Annalisa Crannell, Arnold Feldman, Robert Gethner, Tim Hesterberg, Barbara Nimershiem, George Rosenstein, and William Tyndall—began the process of revising the requirements for a mathematics major. Part of that revision included the development of a course designed to help students make the transition from calculus to upper-level courses by introducing them to conjecture and proof in a context of familiar, easily digestible topics. As such, "Introduction to Higher Mathematics" was born, and I was given the task of teaching it for the first time in the fall of 1995. In the process, I began developing the material that has since evolved into this book.

I particularly wish to acknowledge the contributions of Wendell Culp-Ressler, our resident number theorist, who has served as my sounding board for many of the ideas in this book and who routinely provides me with news on the discovery of the next larger Mersenne prime. He also is the first person, other than the author, to teach out of this text.

One of my students, Benjamin Shanfelder, has been an invaluable asset. Ben was a student in the first offering of "Introduction to Higher Mathematics." In the subsequent years until his graduation in 1998, he served as a teaching assistant for the course. In that role, he had direct contact with the students and could see where they were having difficulties. His observations and insights have led to numerous improvements in this text. He also is largely responsible for writing the solutions manual for the problems at the end of each chapter.

Finally, I'd like to thank Robert Ross, former Acquisitions Editor at Harcourt/ Academic Press, for providing me with several beneficial reviews and for having the faith to go forward with this project.

Tips on Writing Mathematics

One of the major components of this course consists of writing formal proofs and other less formal mathematical arguments. Although some of you may have done some writing in calculus and other courses, this is likely to be the first course you've taken in which writing plays such a prominent role.

Writing in mathematics is, in many ways, like writing in other disciplines—you have to use complete sentences, proper punctuation and grammar, and so on. Aesthetic considerations such as "flowery" language and careful attention to subtle nuances, which are valued in other types of writing, are less important in mathematics. The most important objective in mathematical writing is **clarity**. Remember, you are trying to convince the reader that you know what you're talking about. If you can't express yourself clearly, you'll never achieve that goal.

Here are a few guidelines that might help you:

Audience

You should assume that you are writing for an audience of knowledgeable peers. That means the reader knows at least as much mathematics as you do.

Voice

Mathematics is almost always written in the first person plural ("we," "us," "our"). Do not use "I" or impersonal pronouns such as "one." Try to avoid passive voice also. (This is different from most other sciences, where you would say, for example, "The chemicals were mixed . . . .")

In many cases, personal pronouns can be omitted entirely. For example, say "The roots of the equation are $x = 2$ and $x = 5$," rather than "We found that the roots of the equation are $x = 2$ and $x = 5$." Definitely do not say, "The roots of the equation were found to be $x = 2$ and $x = 5$."

Pay careful attention to dangling participles. It is grammatically incorrect to say, "Using the quadratic formula, the roots of the equation are . . . ." (Or, for that matter, "Digging for a bone, he found his dog.")

Algebraic Details

Given that the reader is reasonably sophisticated mathematically, you can leave out most of the mundane algebraic details. For example, it is perfectly acceptable to say, "The time at which the ball hits the ground is the positive root of the quadratic equation $t^2 + 4t - 5 = 0$, which is $t = 1$," or, "Let $f(x) = x^2 e^x$, whose derivative is $f'(x) = (x^2 + 2x)e^x$." You need not show the calculations.

In some cases, you might wish to provide a description of the technique used, such as, "Using the quadratic formula, we found that the roots were . . . ," or, "After factoring and rearranging terms, we solved for x and y . . . ." In other cases, you may in fact need to include some algebra. If that is the case, think carefully about which steps are necessary to keep the reader on the right track. Do not merely copy the calculations you performed to come up with the answer.

Structure

If a proof or argument is more than a few sentences long, you may wish to state the result first and then a plan or outline of the proof. For example: "We need to show that the function f has a local minimum at $x = 1$. To do so, we'll first show that $f'(1) = 0$. Then we'll show that $f'(x) < 0$ for $x < 1$ and $f'(x) > 0$ for $x > 1$." Then proceed with some of the details.

Do not write a diary. You should not say, "We did this, then we did that. Later, we did the other." Just present the argument, without the "play-by-play."

Connections

Each step in the argument should flow logically from the preceding step. To emphasize the connection between statements, you should freely use words such as *therefore, thus, so, hence, it follows that, consequently, from this,* and so on, where appropriate. They aren't necessary at every step, but should be used at important conclusions. Also, the word *since,* which some disciplines (and editors) consider incorrect, is perfectly acceptable in mathematics. So, for example, you may say, "Since $x > 1$, then $1/x < 1$."

Grammar and Punctuation

Remember that mathematical symbols are "words" and should be used with proper grammar and syntax. Use periods at the ends of sentences, even if the sentences end with mathematical symbols. Use commas, semicolons, colons, and dashes, where appropriate. Do not, however, use exclamation points unless you mean "factorial."

The equal sign ($=$) is a verb and should be used *only* between mathematical symbols. So you could say, "let $x = 2$," but not, "The perimeter of a triangle $=$ the sum of the lengths of the sides."

Example: Here are two versions of the solution to a calculus problem. It should be clear to you which of the two is preferable.

Problem: Determine and classify the critical points of $f(x) = x^3 - 3x$.

Solution 1:
$$f(x) = x^3 - 3x$$
$$f'(x) = 3x^2 - 3 = 0$$
$$3x^2 = 3$$
$$x^2 = 1$$

The critical points are 1 and -1.
$$f''(x) = 6x$$
$$f''(1) = 6\text{—min}$$
$$f''(-1) = -6\text{—max}$$

Solution 2:   Let $f(x) = x^3 - 3x$, whose derivative is $f'(x) = 3x^2 - 3$. Upon setting $f'(x) = 0$, we find that the critical points are $x = 1$ and $x = -1$. The second derivative of f is $f''(x) = 6x$. Since $f''(1) = 6 > 0$, then $x = 1$ is a local minimum. Conversely, since $f''(-1) = -6 < 0$, then $x = -1$ is a local maximum.

Chapter 1
Preliminary Ideas

In this chapter, we introduce you to some of the fundamental concepts that form the basis for the rest of this course. Specifically, you will get your first look at the four basic processes — experimentation, conjecture, proof, and generalization — that are central to creating mathematics. Although we could have chosen any of a number of contexts in which to introduce these processes, we have decided to begin with the integers.

We will investigate many properties of the integers, some of which you will think are obvious, others that are not so obvious. (However, just because a property is obvious does not mean it is unimportant or not worth studying.) Then we'll look at sequences of integers and their properties. Finally, we'll focus on the primes, which in some sense are the building blocks for the integers.

In later chapters, we will use many of the results derived here in a more general context.

1.1. Properties of the Integers

In this section, we will study the *integers*. The integers consist of the *natural* (or *counting*) *numbers*, their negatives, and the number zero. We denote the integers by **Z**. Thus, $\mathbf{Z} = \{\ldots, -3, -2, -1, 0, 1, 2, 3, \ldots\}$.

The integers have several properties that we will assume to be true. First, it seems reasonable to claim that the sum of two integers always results in another integer. More formally, we say that the integers are *closed under addition*.

1. Are the integers closed under multiplication?

2. Are the integers closed under subtraction?

3. Are the integers closed under division?

4. Are the nonnegative integers closed under subtraction?

The other assumptions we'll make are properties of arithmetic operations themselves:

- $a + b = b + a$ and $ab = ba$, for all integers a and b.
- $(a + b) + c = a + (b + c)$ and $(ab)c = a(bc)$, for all integers a, b, and c.
- $a(b + c) = ab + ac$, for all integers a, b, and c.

The first of these is called the *commutative property* of addition and multiplication. The second is called the *associative property*. The third is called the *distributive property* of multiplication over addition. We use these properties so often in ordinary algebra that we don't even realize it. For example, suppose we expand $(a + b)^2 = (a + b)(a + b)$. First we use the distributive property to get $a^2 + ab + ba + b^2$. Then we use the commutative property to write this as $a^2 + ab + ab + b^2 = a^2 + 2ab + b^2$.

In later chapters, we'll look at other sets of numbers — for example, rational numbers and complex numbers — and other sets of mathematical objects such as polynomials. Whatever arithmetic operations we define for those sets will be expressed in terms of the corresponding operations with integers. Consequently, we'll be able to prove whether these sets are closed or not closed under the given operations. We'll also be able to check whether the operations are commutative, associative, and distributive. In this sense, the integers are the "building blocks" of all mathematics.[1]

1.1.1. Even and Odd Integers

The integers can be subdivided into two classes, which we generally call *even* and *odd*. Let's do a simple experiment to generate some conjectures.

5. Add various combinations of even and odd integers and state your results. (Note that it is sufficient to restrict your conjecture to two numbers at a time.)

Your results are actually a proposed *theorem*, which is a mathematical statement that must be proven. Theorems are often expressed in the form "If _____, then _____." For example, you may remember the geometry theorem, "If two sides of a triangle are equal, then the angles opposite those sides are equal."

---

[1] The mathematician Leopold Kronecker once said, "God created the natural numbers, and all the rest is the work of man."

The phrase after the word *if* is called the *antecedent* or *hypothesis*. The phrase after the word *then* is called the *consequent* or *conclusion*.

Sometimes theorems are more conveniently expressed in other forms. For example, consider the theorem "The sum of the angles of a triangle is 180 degrees." This can be expressed also as "If A, B, and C are the angles of a triangle, then $A + B + C = 180$." Note that we introduced *variables* to name the angles. Here, the variables represent the angles in an *arbitrary* triangle (as opposed to one specific triangle).

   6. Restate your results to question 5 as a theorem in if-then form. (Note that there are three cases to consider.)

Theorem 1.1

  (i) If _____, then _____.
  (ii) If _____, then _____.
  (iii) If _____, then _____.

To prove Theorem 1.1, we need precise *definitions* of the terms *even integer* and *odd integer.* Definitions are often stated in one of two forms. For example, we could say either "A triangle is a polygon with three sides" or "If a polygon has three sides, then it is called a triangle."

Let's define an even integer as follows:

Definition x is an even integer if $x = 2n$, where n is an integer.

  7. Write a definition of odd integer.

Note that there is more than one way to define an odd integer. For example, we could say that an odd integer is one that is not even, but that isn't a particularly useful definition for doing proofs. It also assumes that every integer is either even or odd (a fact that we will prove later). To be more useful and precise your definition should look like this: "x is odd if $x = $ _____, where n is an integer."

Theorem 1.1 makes a statement about what happens whenever we add *any* two odd integers, for example. This means that it must be true for *every possible pair* of odd integers. Hence, our proof must be constructed in a way that takes care of all possible cases at one time. Merely demonstrating the truth of the theorem for some *specific* pair(s) of odd integers (e.g., $3 + 5 = 8$) is not sufficient proof.

To further illustrate this idea, let's consider the following statements.

  (i) $x + 3 = 5$.
  (ii) $4(x + 3) = 4x + 12$.

The first statement is true only for one value of x, namely, $x = 2$. Hence, we can say that *there exists an x* such that $x + 3 = 5$. The second statement is true no matter what value of x we substitute. Hence, we can say that *for all x*, $4(x + 3) = 4x + 12$. This is a very important concept. A more detailed discussion of this idea can be found in the Appendix of this book.

On the other hand, if we want to *disprove* a statement, it suffices to provide one *counterexample*. For example, suppose we have the following theorem: "For every real number x, $x^2 \geq x$." This statement is true for most real numbers (e.g., $x = 1$, $x = 3.4$, $x = -2.5$, $x = -0.2$, etc.) However, since $(0.5)^2 = 0.25 < 0.5$, it is not true for all real numbers. Thus, the theorem is false. Sometimes, we can modify a false theorem to make it true. Here, for instance, we could say: "For every real number $x \geq 1$ or $x \leq 0$, $x^2 \geq x$" or "For every *integer x, $x^2 \geq x$*."

To prove Theorem 1.1, we begin by selecting two *arbitrary* odd integers, $x = 2m + 1$ and $y = 2n + 1$.

8. Why didn't we use an "n" for both x and y, as the definition says?

9. Complete the proof of the Theorem 1.1 for all cases.

Using the same procedure, let's try to prove some similar results about multiplication.

10. Repeat question 5 for multiplication instead of addition. Then state and prove a theorem similar to Theorem 1.1.

While we're discussing even and odd integers, here are some other questions for you to consider. Even though the answers may seem obvious, it is important to remember not to jump to conclusions based entirely on observations.

11. Are there any integers that are both even and odd?

You can prove your answer to question 11 by *contradiction*. In other words, assume that there exists some integer x that is both even and odd. Show that this leads to a statement that must be false.

12. Complete the proof above.

13. Is every integer either even or odd? How do you know?

A formal proof of the answer to question 13 requires a theorem called the Division Algorithm, which will be discussed next.

1.1.2. The Division Algorithm

In elementary school, you were probably taught to express the answer to a division problem as a quotient with a remainder; for example,

$$52 \div 6 = 8 \ r \ 4.$$

14. Write a multiplication problem that is equivalent to the division problem above; in other words, write a statement not involving division that you could use to check whether your answer to the division problem is correct.

15. More generally, suppose that when we divide a by b, where a and b are positive integers, we get a quotient q and remainder r. Write the corresponding multiplication statement.

16. Suppose that you had written 52 ÷ 6 = 7 r 10. Your teacher would have marked this wrong, even though the corresponding multiplication statement is correct. Why?

Now let's state a theorem that tells us what happens whenever we divide one positive integer by another. This theorem is known as the *Division Algorithm*. (The word *algorithm* means "mathematical procedure." This theorem is called the Division Algorithm because it tells us what happens when we apply the procedure of division to two integers.)

17. Fill in the blanks in the following theorem.

Theorem 1.2 (Division Algorithm)   Given any two positive integers a and b, there exist unique nonnegative integers q and r such that _____, where _____ $\leq r <$ _____.

The Division Algorithm is an example of an "existence and uniqueness" theorem because it tell us that the integers q and r exist and that there is only one set of values for q and r that satisfy the conditions of the theorem. Theorems of this type are very common in mathematics.

The proof of the "existence" part of this theorem requires an axiom called the Well-Ordering Principle, which states that every nonempty set of positive integers has a smallest element. (This is another one of the properties of the integers that we will assume to be true.) We won't pursue this now; see the Additional Questions at the end of the chapter for a proof. The proof of the "uniqueness" part is outlined below.

Theorem 1.2 also says that q and r exist for every (positive integral) choice of a and b. Thus, when proving the existence part, it is insufficient to show that q and r exist for some specific a and b.

18. Why do we need the last phrase (beginning with "where") in the statement of the Division Algorithm?

19. Use the Division Algorithm to prove that every integer is either even or odd.

When a theorem asserts that some quantity is unique, we need to prove that there is only one such quantity satisfying the conditions of the theorem. Usually we do this as follows:

- Assume that there are two distinct such quantities.
- Show that this assumption leads to a contradiction.

The contradiction leads to the conclusion that there aren't two distinct quantities. Since presumably we've already proved the existence of *at least one* such quantity, then there must be *exactly* one.

In the case of the Division Algorithm, assume that there are two distinct sets of values of q and r that satisfy the theorem. Let's call these values q_1 and r_1, and q_2 and r_2, respectively.

20. Write an equation relating a, b, q_1, and r_1. Write a similar equation relating a, b, q_2, and r_2.

21. Subtract the two equations you wrote in question 20 and rearrange the resulting equation so that all terms with q's in them are on the left side and all terms with r's in them are on the right side.

*22. Carefully argue that the left side of the equation is a multiple of b, but the right side is not. Hence, we have a contradiction. Thus, our initial assumption that there are two distinct sets of values must be incorrect. Since we know there is at least one set of values (why?), then there must be **exactly** one set of values.*

Although the Division Algorithm may seem to be a trivial result hardly worth stating, it is, in fact, a very useful idea that we will see again in later chapters. Moreover, we will encounter other sets of objects that have properties similar to those of the integers. One of the questions we'll ask about these sets is whether the Division Algorithm is true (or whether it needs to be modified) for them.

1.1.3. Congruences

From the Division Algorithm, we now know that every integer is either of the form $2n$ (the even integers) or $2n + 1$ (the odd integers). In other words, we have partitioned the entire set of integers into two disjoint subsets, where the subsets are determined by the remainder when the integer is divided by 2.

At this point, we are ready to generalize. Rather than divide by 2, let's divide by 3 or 4 or any other positive integer m. Clearly, if we divide by m, we will partition the integers into m disjoint subsets.

23. Characterize the subsets into which the integers are partitioned according to the remainder upon division by 3. *List a few integers in each subset. What must be true about the difference between any two integers within the same subset?*

24. Characterize the subsets into which the integers are partitioned according to the remainder upon division by 4. *List a few integers in each subset. What must be true about the difference between any two integers within the same subset?*

25. Characterize the subsets into which the integers are partitioned according to the remainder upon division by m. List a few integers in each subset. What must be true about the difference between any two integers within the same subset?

We're now ready for a definition.

Definition Let a and b be integers and let m be a positive integer. We say that "a is congruent to b mod m," written $a \equiv b$ mod m, if and only if $a - b$ is divisible by m.

In other words, $a \equiv b$ mod m if and only if a and b have the same remainder when divided by m. Or, equivalently, $a \equiv b$ mod m if and only if $a = qm + b$ for some integer q. Note that *mod* is a shortened version of *modulo* or *modulus*.

26. Which of the following are true?
  (i) $8 \equiv 2$ mod 3
  (ii) $17 \equiv -3$ mod 5
  (iii) $43 \equiv 17$ mod 8

27. Determine two negative and two positive integers that are congruent to 3 mod 4.

28. Determine all positive integers m such that $61 \equiv 33$ mod m.

Now we'd like to show how to do arithmetic with congruences. Suppose $a \equiv b$ mod m and $c \equiv d$ mod m. Let's prove that $a + c \equiv (b + d)$ mod m. To get started, we take the given information and write it in an equivalent form:

Since $a \equiv b$ mod m, then $a = qm + b$ for some integer q.

29. What is the next step?

30. Complete the proof.

31. Prove that ac $\equiv$ (bd)mod m.

32. Prove that $a^2 \equiv b^2$ mod m and $a^3 \equiv b^3$ mod m.

33. Is $\dfrac{a}{c} \equiv \dfrac{b}{d}$ mod m?

The result of question 32 can be extended to all positive integer exponents. Proving this requires mathematical induction, which we will learn in Section 1.3.

The idea of congruences is very useful. For example, suppose we want to prove that $n^5 - n$ is divisible by 5 for all integers n. One way to prove this is to look at five cases, one for each subset into which the integers are partitioned modulo 5. Here are three of the five cases:

> *Case 1:* Suppose $n \equiv 0$ mod 5. Then $n^5 - n \equiv (0^5 - 0)$ mod $5 \equiv 0$ mod 5. So, $n^5 - n$ is divisible by 5.
>
> *Case 2:* Suppose $n \equiv 1$ mod 5. Then $n^5 - n \equiv (1^5 - 1)$ mod $5 \equiv 0$ mod 5. So, $n^5 - n$ is divisible by 5.
>
> *Case 3:* Suppose $n \equiv 2$ mod 5. Then $n^5 - n \equiv 30$ mod $5 \equiv 0$ mod 5. So, $n^5 - n$ is divisible by 5.

34. Complete the remaining two cases.

Since we have accounted for all possible values of n, the proof is complete. (We'll see another way to prove this statement later.)

We're now going to look at an example of an extremely common theme in mathematics. We'll define a set of objects (in this case, numbers) and some operations on those objects. Then we'll explore some interesting properties of that set under the given operations.

Let $Z_m = \{0, 1, 2, \ldots, m - 1\}$. For any two elements x and y of Z_m, we define $x + y$ by taking the "ordinary" sum of x and y and "reducing" it, modulo m, to an element of Z_m. For example, $Z_6 = \{0, 1, 2, 3, 4, 5\}$. Then $3 + 5 = 2$ since, in ordinary arithmetic, $3 + 5 = 8$ and $8 \equiv 2$ mod 6. Multiplication is defined in a similar manner. Hence, in Z_6, $3(5) = 3$, $3(4) = 0$, and $2(5) = 4$.

Another way to think of this is that each element x of Z_m represents an entire subset of the set of integers, namely, those integers congruent to x mod m. When we add (or multiply) two elements of Z_m, we are really adding (multiplying) any two elements of the corresponding subsets. So the fact that $3 + 5 = 2$ in Z_6 means that when we add *any* integer congruent to 3 mod 6 and *any integer congruent to 5 mod 6, we get an integer congruent to 2 mod 6.*

35. Complete the following addition table for Z_4.

+	0	1	2	3
0	0	1	2	3
1	1	2	3	0
2				
3				

36. *Create a multiplication table for Z_4.*

37. *Create addition and multiplication tables for Z_5 and Z_6.*

38. *Think of a practical situation in which we do arithmetic in Z_{12}.*

You should have noticed at least one unusual phenomenon. In Z_4 and Z_6, it is possible to have the product of two nonzero elements equal to zero. This does not happen in the entire set of integers.

39. *Does it happen in Z_5?*

40. *In which Z_m would you expect to find the product of two nonzero elements equal to zero?*

1.1.4. Divisibility

Integers are numbers such as three, fourteen, two hundred fifty-six, and so on. We express these numbers as decimal (base 10) numerals: 3, 14, 256. Notice that the value of the numeral 256 is $2(10)^2 + 5(10) + 6$. More generally, an integer N can be expressed as the numeral $a_r a_{r-1} \ldots a_2 a_1 a_0$. So, for example, in the numeral 256, $a_0 = 6$, $a_1 = 5$, and $a_2 = 2$.

41. *What is the value of the integer N whose decimal numeral is $a_r a_{r-1} \cdots a_2 a_1 a_0$?*

42. *Prove that $N \equiv a_0 \bmod 10$.*

43. *How can you tell by looking at its numeral if N is divisible by 2? by 5?*

44. *How can you tell by looking at its numeral if N is divisible by 4? by 8? Generalize.*

45. *Prove that $N \equiv a_0 + a_1 + \cdots + a_r) \bmod 9$. How can you use this fact to determine whether N is divisible by 9?*

46. *Devise a criterion similar to the one above for divisibility by 11.*

1.2. Conjectures and Proofs About Sequences

As we said earlier, one of the main themes of this course involves performing experiments to generate data from which conjectures can be formulated. Then we try to prove the conjecture.

In many respects, conjecturing is the most difficult aspect of this course. Do not be discouraged if you struggle with it at first; you'll have ample opportunity for more practice as you proceed through the remaining chapters.

One area that is especially good for generating conjectures involves sequences of numbers. In general, let $\{u_1, u_2, u_3, \ldots\}$ be a sequence of numbers. In some cases, we are given an explicit formula for the jth term of the sequence in terms of j.

1. *Write the first five terms of the sequence defined by* $u_j = 3j + 1$.

2. *Write the first five terms of the sequence defined by* $u_j = 2 \cdot 3^{j-1}$.

3. *Write a formula for* u_j *in each of the following sequences.*

   (i) $\{1, 5, 9, 13, 17, \ldots\}$
   (ii) $\{2, 5, 10, 17, 26, \ldots\}$
   (iii) $\left\{ \dfrac{1}{1}, \dfrac{2}{3}, \dfrac{3}{5}, \dfrac{4}{7}, \ldots \right\}$
   (iv) $\{1, 3, 7, 15, 31, \ldots\}$

In other cases, we may be given a *recursive* formula — that is, a formula that expresses the jth term of the sequence as a function of one or more previous terms. For example, suppose $u_1 = 3$ and $u_j = u_{j-1} + 2$ for $j \geq 2$.

4. *Write the first five terms of this sequence.*

5. *Write the first five terms of the sequence defined by* $u_1 = 1$, $u_2 = 3$, *and* $u_j = 2u_{j-1} + u_{j-2}$ *for* $j \geq 3$.

6. *Determine a recursive formula for the sequences in questions 1 and 2.*

In general, converting between explicit and recursive formulas for sequences is not easy, and we shall not attempt to do so except in special cases. However, it is possible to determine whether a given explicit formula generates the same sequence as a given recursive formula.

For example, consider the sequence defined recursively by $u_1 = 2$ and $u_j = u_{j-1} + 2j$, for $j \geq 2$. Then consider the explicit formula $u_j = j(j + 1)$. We could generate the first few terms for each formula and show that they agree. However, this does not guarantee that they will agree for all j. To do that, we need a proof. The easiest way to prove that they generate the same sequence is to substitute the explicit formula into the recursive formula and show that a true statement results. (More formally, we need to use mathematical induction, a technique that will be covered in the next section.)

7. *If* $u_j = j(j + 1)$, *what is* u_{j-1}?

8. *Substitute* u_{j-1} *into the recursive formula, simplify, and show that you get* u_j. *Since both formulas give the same* u_1, *the proof is complete.*

We'll also be dealing with summations of terms in a sequence. Such summations are called *series*. We'll represent them using *sigma notation*. The sum of the first n terms of the sequence u_j is

$$S_n = u_1 + u_2 + \cdots + u_n = \sum_{j=1}^{n} u_j.$$

9. *Represent each of the following in sigma notation.*
   (i) *The sum of the first n odd-numbered terms; i.e.,* $u_1 + u_3 + \cdots + u_{2n-1}$.
   (ii) *The sum of the first n even-numbered terms; i.e.,* $u_2 + u_4 + \cdots + u_{2n}$
   (iii) *The alternating sum of the first n terms; i.e.,* $u_1 - u_2 + u_3 - \cdots \pm u_n$.

10. *Express each of the following in sigma notation:*
    (i) $1^2 + 2^2 + 3^2 + \cdots + n^2$
    (ii) $\dfrac{1}{1(2)} + \dfrac{1}{2(3)} + \dfrac{1}{3(4)} + \cdots + \dfrac{1}{n(n+1)}$
    (iii) $\dfrac{1}{1!} - \dfrac{1}{3!} + \dfrac{1}{5!} - \cdots \pm \dfrac{1}{(2n+1)!}$

Recall that the notation $n!$, read "n factorial," represents the product of the first n positive integers. Thus, $3! = 3(2)(1) = 6$, $5! = 5(4)(3)(2)(1) = 120$, and so on. By convention, we define $0!$ to be 1.

Now we'll look at some of the aforementioned special cases.

1.2.1. Arithmetic and Geometric Sequences

Let $S_n = \sum_{j=1}^{n} (2j - 1)$ be the sum of the first n terms of the sequence $u_j = 2j - 1$.

11. *Evaluate S_n for several values of n and propose a formula for S_n.*

To prove this formula, note that $S_n = 1 + 3 + 5 + \cdots + (2n - 1)$. By reversing the order of the terms, we get $S_n = (2n - 1) + (2n - 3) + \cdots + 3 + 1$.

12. *Add these two expressions for S_n, noting that the sum of the two first terms is $(2n - 1) + 1 = 2n$, the sum of the two second terms is $(2n - 3) + 3 = 2n$, and so on. Then solve for S_n.*

13. *Use a similar approach to derive formulas for $\sum_{j=1}^{n} 2j$ and $\sum_{j=1}^{n}(3j - 1)$.*

14. Can the same approach be used to derive a formula for $\sum_{j=1}^{n} j^2$?

Sequences of the form $u_j = a + b(j - 1), j = 1, 2, 3, \ldots$, where a and b are constants, are called *arithmetic sequences*.

15. Determine a recursive formula for an arithmetic sequence.

16. Complete and prove the following theorem.

Theorem 1.3   The sum of the first n terms of an arithmetic sequence is given by

$$S_n = \sum_{j=1}^{n} a + b(j - 1) = \underline{\hspace{3cm}}.^2$$

Now let $S_n = \sum_{j=1}^{n} 2^{j-1}$ be the sum of the first n terms of the sequence $u_j = 2^{j-1}$.

17. Propose a formula for S_n.

18. To prove your result, note that $S_n = 1 + 2 + 4 + \cdots + 2^{n-1}$. Upon multiplying by 2, we get $2S_n = 2 + 4 + \cdots + 2^n$. Subtract these two equations and solve for S_n.

19. Use a similar approach to find a formula for $S_n = \sum_{j=1}^{n} 3^{j-1}$.

Sequences of the form $u_j = ar^{j-1}, j = 1, 2, 3, \ldots$, where a and r are constants and with $r \neq 0$ or 1, are called *geometric sequences*.

20. Write a recursive definition of a geometric sequence.

21. Complete and prove the following theorem.

Theorem 1.4   The sum of the first n terms of a geometric sequence is given by

$$S_n = \sum_{j=1}^{n} ar^{j-1} = \underline{\hspace{3cm}}.$$

22. What happens to the geometric sequence and its corresponding series if $r = 1$?

---

[2] There is a story that when he was a young lad, the mathematician Karl Gauss was told to add the integers from 1 to 1000. Allegedly, this assignment was supposed to keep the rowdy young Karl quiet for a while. Much to his teacher's chagrin, Karl arrived at the answer very quickly because, in essence, he had discovered this theorem or, at least, the idea of "adding from both ends toward the middle."

Theorem 1.4 gives us the sum of a finite number of terms in a geometric sequence. In certain cases, it is possible to add an infinite number of terms and obtain a finite sum. To do this, we first need to recall from calculus what is meant by adding an infinite number of terms.

Definition   Let $\{u_1, u_2, \ldots\}$ be a sequence of numbers and let $S_n = \sum_{j=1}^{n} u_j$ be the sum of the first n terms of the sequence. If $\lim_{n\to\infty} S_n$ exists, then we say that the *infinite series* $\sum_{j=1}^{n} u_j$ *converges*, and we define $\sum_{j=1}^{\infty} u_j = \lim_{n\to\infty} S_n$.

In other words, if we want to add up infinitely many numbers in a sequence, we add up the first n terms and take the limit as n approaches ∞. If the limit exists, then the infinite series converges to whatever that limit is. If the limit does not exist, then the infinite series diverges.

At this point, we are relying on an intuitive notion of what it means for a sequence to have a limit. Suppose we claim $\lim_{n\to\infty} S_n = L$. Roughly speaking, this means that we have to be able to find a point in the sequence beyond which all terms are arbitrarily close to L. For example, let

$$S_n = \frac{n}{n+1} = \left\{ \frac{1}{2}, \frac{2}{3}, \frac{3}{4}, \ldots \right\}.$$

We claim $\lim_{n\to\infty} S_n = 1$. The reason this is correct is as follows: All terms beyond S_{100} are within 0.01 of 1; all terms beyond S_{1000} are within 0.001 of 1. Indeed, whatever tolerance we choose, we can find a point beyond which all terms are that close or closer to 1. On the other hand, had we claimed $\lim_{n\to\infty} S_n = 2$, we would be wrong because there are no terms within 0.01 of 2, or 0.1 of 2, or even 0.9 of 2. Also, $\lim_{n\to\infty} S_n$ is not 0.75 because, even though there is a term that is arbitrarily close to 0.75 (in fact, equal to 0.75), the succeeding terms get farther away from 0.75.

We will encounter the notion of limit of a sequence again in Chapter 5. A more detailed approach can be found in many standard college calculus or analysis textbooks.

In the case of the geometric sequence, Theorem 1.4 gives us an expression for S_n.

23. *Argue that if* $-1 < r < 1$, *then* $\lim_{n\to\infty} S_n = \dfrac{a}{1-r}$.

Thus, we have the following corollary.

Corollary 1.1   If $-1 < r < 1$, then $\displaystyle\sum_{j=1}^{\infty} ar^{j-1} = \frac{a}{1-r}$.

24. *Evaluate each of the following.*

(i) $1 + \dfrac{1}{3} + \dfrac{1}{9} + \dfrac{1}{27} + \cdots$

(ii) $1 - \dfrac{1}{2} + \dfrac{1}{4} - \dfrac{1}{8} + \cdots$

(iii) $0.999\ldots = \dfrac{9}{10} + \dfrac{9}{100} + \dfrac{9}{1000} + \cdots$

1.2.2. Fibonacci Sequence

There is a well-known sequence of integers that has many interesting properties. Discovering these properties is a particularly good exercise in generating hypotheses.

Consider the sequence $\{1, 1, 2, 3, 5, 8, 13, 21, 34, \ldots\}$. This sequence is called the *Fibonacci*[3] *sequence.*

25. *Let f_j be the jth term of the sequence. Write a recursive definition of the sequence and use it to compute the next five terms.*

26. *Propose an expression for each of the following summations of the Fibonacci sequence. Your answers will not be explicit functions of n, but rather will be functions of other terms of the Fibonacci sequence.*
    (i) *The sum of the first n terms.*
    (ii) *The sum of the first n odd-numbered terms.*
    (iii) *The sum of the first n even-numbered terms.*
    (iv) *The sum of the squares of the first n terms.*

27. *To prove your result in 26(i), note that, by the definition of the sequence,*
    $f_1 + f_2 + f_3 + f_4 + \cdots = (f_3 - f_2) + (f_4 - f_3) + (f_5 - f_4) + \cdots.$
    Complete the proof.

28. *Provide similar proofs for 26(ii) and 26(iii).*

29. *For 26(iv), note that we can rewrite the sum of squares as $f_1^2 + f_2^2 + f_3^2 + \cdots = f_1^2 + f_2(f_3 - f_1) + f_3(f_4 - f_2) + \cdots$. With a little careful bookkeeping, you can complete the proof. Do so.*

The Fibonacci sequence has many other interesting properties. Here are a few for you to explore. We'll defer the proofs for now.

30. *Notice that $5^2 = 3(8) + 1$, $8^2 = 5(13) - 1$, $13^2 = 8(21) + 1$, $21^2 = 13(34) - 1$, and so on. Express this pattern in general.*

---

[3] The Fibonacci sequence was discovered by Leonardo of Pisa in the 13th century. (He was the son [*fils*, in French] of Bonaccio, hence the name of the sequence.) It has been studied at great length. In fact, there is a journal, called the *Fibonacci Quarterly*, devoted to this and related sequences.

31. The pattern above relates the square of the nth term to the product of the (n + 1)st and (n − 1)st terms. Find a similar pattern relating the square of the nth term to the product of the (n + 2)nd and (n − 2)nd terms.

32. More generally, find a pattern relating the square of the nth term to the product of the (n + r)th and (n − r)th terms for any r.

33. Notice that 5(8) = 3(13) + 1 = 2(21) − 2 = 1(34) + 6. See if you can find a general pattern of this type.

We have defined the Fibonacci sequence recursively. It is possible to give an explicit formula, although it is quite complex and not at all readily apparent. We'll investigate that in Chapter 5.

1.3. Mathematical Induction

In the previous section, we proposed a number of statements that were true for all positive integer values of n. For example, $\sum_{j=1}^{n}(2j - 1) = n^2$ or $\sum_{j=1}^{n}2^{j-1} = 2^n - 1$. We proved these statements *deductively* — that is, by writing a sequence of statements, each of which flowed logically from the previous one. An advantage to this approach, at least in the examples we've seen so far, is that we could use the proof to actually derive the result. In other words, we didn't need to know the formula to prove it; following the same steps would produce the formula for us. A disadvantage to this method is that, in some cases, it requires a bit of ingenuity (perhaps motivated by having conjectured the result we're trying to derive) to get started.

There is another type of proof that can be used to prove that a given statement is true for all positive integer values of n. So, for example, the statement might be a formula like $\sum_{j=1}^{n}2^{j-1} = 2^n - 1$ or it might be a statement like "$n^3 - n$ is divisible by 3." This method is called *mathematical induction* and is based on the following.

Principle of Mathematical Induction   Any set of positive integers containing the integer 1 and containing the integer $k + 1$ whenever it contains k must contain all positive integers.

The Principle of Mathematical Induction is an axiom or assumption about the positive integers; thus, it cannot be proven. It is, however, equivalent to the Well-Ordering Principle mentioned in Section 1.1.

Invoking the Principle of Mathematical Induction is a two-step process:

Step 1: Show that the statement is true for $n = 1$.

Step 2: Assume that the statement is true for $n = k$ and show, based on this assumption, that it is true for $n = k + 1$.

The assumption in Step 2 is called the *induction hypothesis*.

To better understand the Principle of Mathematical Induction, imagine that you have set up a series of equally spaced domino blocks. The first part of the induction procedure implies that the first block falls. The second part says that the blocks are close enough together so that whenever one block falls, the next one falls. Since the first one falls, then the second one falls, which, in turn, means that the third one falls, and so on. Hence, they all fall.

1. *How would you modify the induction procedure if you wanted to show the statement is true for all odd positive integers? for all even positive integers?*

2. *How would you modify the procedure if you wanted to show the statement is true for all integers greater than or equal to 10?*

Let's try induction to prove $\sum_{j=1}^{n} 2^{j-1} = 2^n - 1$. The statement is true for $n = 1$ since $\sum_{j=1}^{1} 2^{j-1} = 2^0 = 1 = 2^1 - 1$. Now assume that the statement is true for $n = k$; that is, $\sum_{j=1}^{k} 2^{j-1} = 2^k - 1$. We want to show that the statement is true for $n = k + 1$; that is, $\sum_{j=1}^{k+1} 2^{j-1} = 2^{k+1} - 1$. To do so, we note that $\sum_{j=1}^{k+1} 2^{j-1} = \sum_{j=1}^{k} 2^{j-1} + 2^k$. We use the induction hypothesis to replace the first term on the right side of the last equation. Thus,

$$\sum_{j=1}^{k+1} 2^{j-1} = 2^k - 1 + 2^k = 2(2^k) - 1 = 2^{k+1} - 1$$

and the proof is complete.

Now let's try to prove that $n^3 - n$ is divisible by 3 for all n.

3. *Is the statement true when $n = 1$?*

4. *What is the induction hypothesis? What do you want to show?*

5. *Complete the proof. [Hint: $(k + 1)^3 - (k + 1) = (k^3 - k) + 3k^2 + 3k$.]*

6. *Use induction to prove each of the following.*
   (i) $\sum_{j=1}^{n} (2j - 1) = n^2$
   (ii) $\sum_{j=1}^{n} \frac{1}{j(j + 1)} = \frac{n}{n + 1}$
   (iii) $\sum_{j=1}^{n} j^2 = \frac{n(n + 1)(2n + 1)}{6}$
   (iv) $\sum_{j=1}^{n} f_j = f_{n+2} - 1$, *where f_j is the Fibonacci sequence*
   (v) $\sum_{j=1}^{n} f_{2j-1} = f_{2n}$
   (vi) $4^n - 1$ *is divisible by 3 for all n.*

7. *Is it true that $n^4 - n$ is divisible by 4 for all n? If not, show where the induction proof fails.*

1.4. Primes and the Fundamental Theorem of Arithmetic

In this section, we'll investigate some more properties of the integers. In particular, we'll emphasize prime numbers and the factoring of integers as a product of primes. Like the Division Algorithm, primes and factoring are concepts that apply to other types of mathematical objects such as polynomials. Recall from elementary algebra that you can factor $x^2 - 4x - 5$ as $(x - 5)(x + 1)$. You cannot factor $7x - 9$ so, in some sense, $7x - 9$ is prime. (We'll explore this in more detail in Chapter 3.)

1.4.1. Prime Numbers

We begin by exploring some properties of prime numbers.

1. *Write a definition of prime number.*

2. *Make a list of the first 20 prime numbers.*

3. *Is 1 a prime?*

An interesting, but surprisingly difficult, problem is determining whether a given integer is prime. One way to do this is to just divide the integer by successive primes and see whether any of the remainders are 0. If the given integer is large, this could be a tedious process, even with a computer.

4. *If you wanted to detemine whether an integer N is prime, what is the largest prime by which you would have to divide before you could come to a conclusion? Which of the following are prime?*
    (i) 173
    (ii) 323
    (iii) 1073
    (iv) 4891

There are some "bookkeeping" methods of determining all the primes less than or equal to some number. One such method is called the *Sieve of Eratosthenes*. In the table below, first cross out all multiples of 2. Then cross out all multiples of 3, of 5, of 7, and of 11. The uncrossed numbers are the primes.

	2	3	4	5	6	7	8	9	10
11	12	13	14	15	16	17	18	19	20
21	22	23	24	25	26	27	28	29	30
31	32	33	34	35	36	37	38	39	40
41	42	43	44	45	46	47	48	49	50
51	52	53	54	55	56	57	58	59	60
61	62	63	64	65	66	67	68	69	70
71	72	73	74	75	76	77	78	79	80
81	82	83	84	85	86	87	88	89	90
91	92	93	94	95	96	97	98	99	100

Next we'd like to determine whether there is some largest prime integer beyond which all integers are composite. In other words, are there finitely many primes? The answer to this question is no, as the next theorem states.

Theorem 1.5   There are infinitely many primes.

The proof (due to Euclid, c. 300 B.C.) is by contradiction and is outlined below.

5. *Suppose there are only finitely many primes, $p_1, p_2, p_3, \ldots, p_r$. Let $N = p_1 p_2 p_3 \cdots p_r + 1$. Clearly, $N > p_i$ for all i. We claim N must be composite. Why?*

6. *If N is composite, then it must be divisible by one of the primes $p_1, p_2, p_3, \ldots, p_r$. But N is not divisible by any of the primes. Why not?*

7. *Therefore, N must be prime. Why is this a contradiction?*

Hence, our initial assumption that there are only finitely many primes is wrong. So there must be infinitely many primes.

We've just shown that there are infinitely many primes. However, now we'll show that there are arbitrarily long sequences of consecutive composite numbers. In other words, it is possible to find one million consecutive composite numbers. Or one billion, or one trillion, etc. Yet the primes never stop.

8. *Find 5 consecutive composite numbers. Find 10 consecutive composite numbers.*

9. *Show that the numbers 6! + 2, 6! + 3, 6! + 4, 6! + 5, and 6! + 6 are all composite.*

10. *Suggest a scheme for finding k consecutive composite numbers, for any integer k.*

11. *Is this a contradiction to the fact that there are infinitely many primes?*

In case you aren't already confused, it is also possible to show that the relative frequency of primes decreases. For example, there are 25 primes less than 100, for a relative frequency of .25. There are 168 primes less than 1000, for a relative frequency of .168. There are 664,579 primes less than 10,000,000, for a relative frequency of .066. So, in some sense, the primes become scarce, but they never run out. There is a theorem, called the *Prime Number Theorem*, that characterizes the fraction of primes less than n, for large values of n.[4] Its proof is well beyond the scope of this text, so we won't pursue it here.

Now let's look at a different problem. Rather than trying to determine which integers are prime, let's try to eliminate as many possible primes as we can. Obviously, no even integer greater than 2 can be prime. The question is: Can we do better than this; that is, can we eliminate more integers from the list of potential primes?

12. Divide each of the first 20 *primes by* 6 *and note the remainder.*

You should have noticed that, except for $p = 2$ or 3, all of the remainders are either 1 or 5. Thus, we have the following theorem.

Theorem 1.6 If p is a prime greater than 3, then $p = 6n + 1$ or $p = 6n + 5$, where n is an integer.

13. Restate Theorem 1.6 in terms of congruences.

To prove Theorem 1.6, we'll use the Division Algorithm. Specifically, every integer, upon division by 6, leaves a remainder r, where $0 \leq r \leq 5$. So, we need to consider six cases, one for each possible remainder.

14. Show that if the remainder is 0, 2, 3, *or* 4, *then p must be composite.*

Theorem 1.6 further limits the possibilities for primes by eliminating all those numbers that don't leave remainders of 1 or 5 upon division by 6.

Notice that the theorem does not say that if $p = 6n + 1$ or $p = 6n + 5$, then p is prime. However, it is true that if p is not of the form $6n + 1$ or $6n + 5$, then p is not prime.

15. Is the statement "if p = 6n + 1 *or p* = 6n + 5, *then p is prime" true?*

As we stated in Section 1.1, many theorems can be written in the form, "If P, then Q," where P (the hypothesis) and Q (the conclusion) are some statements. It is important to realize that this is **not** the same as "If Q, then P."

---

[4] The theorem says that $\pi(n) \sim n/\ln(n)$, where $\pi(n)$ is the fraction of integers less than n that are prime. In other words, the ratio of $\pi(n)$ to $n/\ln(n)$ approaches 1 as n approaches ∞. The theorem was first proposed by Gauss in the early 19th century. It was finally proved in 1896 by both Jacques Hadamard and Charles de la Vallée Poussin, who, curiously, arrived at the proof independently and nearly simultaneously.

There are other ways of saying, "If *P*, then *Q*." Among them are:

- *P* only if *Q*.
- *P* is sufficient for *Q*.
- *Q* is necessary for *P*.

The statement obtained by interchanging the hypothesis and conclusion is called the *converse* of the theorem. The fact that a theorem is true says nothing about the truth of its converse and, in fact, in many cases, the converse is false. If the converse happens to be true, then statements *P* and *Q* are *equivalent,* and we can write the combined theorem as "*P* if and only if *Q*" (abbreviated "*P* iff *Q*"). Or, we can say, "*P* is necessary and sufficient for *Q*."

The statement obtained by negating the hypothesis and conclusion and then interchanging them — that is, "If not *Q*, then not *P*" — is called the *contrapositive* of the theorem. The contrapositive of a theorem is always true if the theorem itself is true.

Finally, the statement obtained simply by negating the hypothesis and conclusion — that is, "If not *P*, then not *Q*" — is called the *inverse* of the theorem. Like the converse, the inverse need not be true if the theorem is true. However, if the converse is true, then the inverse is true. (Note that the inverse is the contrapositive of the converse.) A more formal discussion of if-then statements can be found in the Appendix.

We sometimes prove a theorem by proving its contrapositive. This is, in essence, what we did in question 14. We showed that if *p* is not of the form $6n + 1$ or $6n + 5$, then *p* is composite (not prime).

1.4.2. The Fundamental Theorem of Arithmetic

In this section, we will investigate a fact about the positive integers that at first glance may seem obvious and not requiring proof. In actuality, this fact depends on some rather subtle properties of the integers. We'll see how this works as we go through the proof. This result is known as the Fundamental Theorem of Arithmetic. The next few questions illustrate the main ideas behind this theorem.

16. *Express each of the following numbers as a product of prime numbers. Are your answers unique?*
    (i) 60
    (ii) 154
    (iii) 504
    (iv) 1001
    (v) 61,308

17. *Can every positive integer be expressed uniquely as a product of primes, assuming we ignore the order in which the prime factors are written?*

Now we're ready to state the theorem.

18. *Fill in the blanks in the theorem below.*

Theorem 1.7 (Fundamental Theorem of Arithmetic) Let p_1, p_2, p_3, ... be primes. Then for every integer $n \geq 2$, there exist nonnegative integers a_1, a_2, a_3, ... such that $n = $ _____. Moreover, this representation is unique except _____.

Note: When we express n in this form, we say that we have written its *canonical representation*.

This is another existence and uniqueness theorem and, as such, its proof has two parts. First we have to show that every integer has such a representation as a product of primes. Then we have to show that the representation is unique.

We'll prove the "existence" part by using a slightly different form of mathematical induction. Specifically, we'll first show that the statement is true for $n = 2$. Then we'll assume the statement is true for all *n less than or equal to some number k* (this is now the induction hypothesis), and show that it is true for $n = k + 1$.

19. *Why is the statement true for $n = 2$?*

20. *What does it mean, in the context of the theorem, to assume the statement is true for all $n \leq k$?*

21. *To show that the statement is true for $n = k + 1$, we need to consider two cases: If $k + 1$ is prime, there is nothing to show. Why?*

22. *If $k + 1$ is composite, then by definition, $k + 1 = rs$, where r and s are integers less than or equal to k. Now use the induction hypothesis to complete the proof.*

Now we'll show that the representation as a product of primes is unique. As in Section 1.1, we prove uniqueness by assuming there are two such representations and showing they must be the same. First, we need a preliminary result, or *lemma*. This lemma is the property of the positive integers that makes the Fundamental Theorem true.

Lemma 1.1 If b and c are integers and p is a prime such that bc is divisible by p, then either b is divisible by p or c is divisible by p.

Although we could prove this lemma, we won't do so here. However, you should convince yourself that the lemma is plausible by constructing some examples. In particular, why does p need to be prime?

23. Assume $n = p_1p_2p_3 \ldots p_r = q_1q_2q_3 \ldots q_s$, where the p's and q's are primes. We can assume that none of the p's are equal to any of the q's. Why?

24. Clearly, $p_1p_2p_3 \ldots p_r$ is divisible by p_1. Therefore, $q_1q_2q_3 \ldots q_s$ must be divisible by p_1. This, in turn, means that one of the q's must be divisible by p_1. Why? (More formally, q_j is divisible by p_1 for some j.)

25. It follows that $p_1 = q_j$ for some j. Why?

26. Where is the contradiction?

To convince you that the Fundamental Theorem of Arithmetic is not as obvious as it first may have appeared, we conclude this section with the following example. Consider the set S of integers that are congruent to 1 mod 3; that is,

$$S = \{1, 4, 7, 10, 13, \ldots\}.$$

27. Let the "primes" be any numbers in S that are not the product of any two numbers in S other than 1 and itself. Write the first 10 primes in S.

28. Find a number in S that can be written as a product of primes in two distinct ways.

The fact that there are numbers in S that can be written as products of primes in two distinct ways does not contradict the Fundamental Theorem of Arithmetic, which applies to the entire set of positive integers. The set of positive integers has a property (called the *unique factorization property*) that S does not have. In particular, Lemma 1.1 is not true for S.

29. Find a counterexample; that is, find two numbers in S whose product is divisible by some "prime" p, yet neither of the numbers is divisible by p.

As we progress through the course, we will encounter other sets of numbers or, more generally, sets of mathematical objects. Where appropriate, we'll ask whether these sets have the unique factorization property.

1.4.3. Divisors of an Integer and Perfect Numbers

The Fundamental Theorem of Arithmetic has many corollaries. One of them allows us to count the number of distinct divisors of a given integer.

30. Select several positive integers. For each, write its canonical representation. Then make a list of all its divisors and their canonical representations.

How are the canonical representations of the divisors related to that of the integer itself?

The next question is whether we can determine the number of divisors of an integer by looking at its canonical representation. Let's first consider the special case in which $n = p^a$, for some integer a.

31. *How many distinct divisors does n have? What are they?*

32. *Suppose $n = p_1^{a_1} p_2^{a_2}$. How many divisors does n have?*

More generally, suppose $n = p_1^{a_1} p_2^{a_2} p_3^{a_3} \ldots p_r^{a_r}$ and m is a divisor of n.

33. *What must be true about the canonical representation of m?*

The number of divisors of a positive integer n is usually denoted $\tau(n)$.

34. *Complete the following theorem:*

Theorem 1.8 Let $n = p_1^{a_1} p_2^{a_2} p_3^{a_3} \ldots p_r^{a_r}$. The number of divisors of n is given by

$$\tau(n) = \underline{\hspace{5cm}} .$$

In addition to finding the number of divisors of a positive integer, it is also possible to find the *sum* of the divisors, denoted $\sigma(n)$. If $n = p^a$, then clearly $\sigma(n) = 1 + p + p^2 + \cdots + p^a$.

35. *Suppose $n = p_1^{a_1} p_2^{a_2}$. Using your answer to question 32, show that*

$$\sigma(n) = (1 + p_1 + p_1^2 + \cdots + p_1^{a_1})(1 + p_2 + p_2^2 + \cdots + p_2^{a_2}).$$

36. *Generalize to the case in which $n = p_1^{a_1} p_2^{a_2} p_3^{a_3} \ldots p_r^{a_r}$ and complete the following theorem.*

Theorem 1.9 Let $n = p_1^{a_1} p_2^{a_2} p_3^{a_3} \ldots p_r^{a_r}$. The sum of the divisors of n is given by

$$\sigma(n) = \underline{\hspace{5cm}} .$$

Recall from Section 1.2 that the sum of a finite geometric sequence is given by

$$\sum_{j=1}^{n} ar^{j-1} = \frac{a(r^n - 1)}{r - 1}, \text{ if } r \neq 1.$$

37. Use this fact to show that

$$\sigma(n) = \prod_{j=1}^{r} \frac{p_j^{a_j+1} - 1}{p_j - 1}.$$

Note: The capital pi, Π, in question 37 is similar to sigma notation, except that you *multiply* successive terms rather than add them.

Definition   A number is said to be *perfect* if $\sigma(n) = 2n$. In other words, *n* is perfect if the sum of its proper divisors (those divisors that are strictly less than *n*) is equal to *n*.

38. Show that 6, 28, and 496 are perfect. (In fact, they are the three smallest perfect numbers.)

39. It can be shown that if $2^m - 1$ is prime, then $2^{m-1}(2^m - 1)$ is a perfect number. Show that 6, 28, and 496 are of this form.

40. Find the next perfect number of this form.

Next we'd like to show that every number of this form is perfect. Suppose $n = 2^{m-1}(2^m - 1)$, where $2^m - 1$ is prime.

41. List all the divisors of n.

42. Compute the sum of the divisors of n and thus show that n is perfect.

It can also be shown that every *even* perfect number is of the form $2^{m-1}(2^m - 1)$, where $2^m - 1$ is prime. It is not known whether there are any odd perfect numbers, although there is overwhelming evidence that there are not.

Not every number of the form $2^m - 1$ is prime. (Try $m = 4$.) An interesting challenge is to determine those values of *m* for which $2^m - 1$ *is* prime. Primes of this form are called *Mersenne* primes.[5]

43. Show that if $a \neq 2$, then $a^m - 1$ is composite for all m.

44. Show that if m is composite, then $2^m - 1$ is composite.

Thus, we need only consider prime values of *m* when searching for Mersenne primes. However, not every prime value of *m* yields a Mersenne prime.

45. Determine the smallest prime value of m for which $2^m - 1$ is composite.

---

[5] Father Marin Mersenne (1588–1648) was a French monk. He claimed in 1644 that $2^p - 1$ is prime for $p = 2, 3, 5, 7, 13, 17, 19, 31, 67, 127, 257$ and composite for all other primes less than 257. Of course, he didn't have the computing machinery to check this, and it turns out he was wrong. In fact, using modern computers, it has been shown that $2^{67} - 1$ and $2^{257} - 1$ are composite, while $2^{61} - 1$, $2^{89} - 1$, and $2^{107} - 1$ are prime.

Note: The problem of determining which prime values of m result in Mersenne primes is as yet unsolved. The largest known Mersenne prime is $2^{6,972,593} - 1$, which is a number with 2,098,960 digits.

1.4.4. Greatest Common Divisor and Least Common Multiple

The Fundamental Theorem of Arithmetic can be used to investigate the greatest common divisor and least common multiple of two integers.

Definition The *least common multiple* of m and n, denoted lcm(m, n), is the smallest number u such that u is divisible by both m and n.

46. *Determine the least common multiple of each of the following.*
    (i) 8 *and* 12
    (ii) 84 *and* 132
    (iii) 108 *and* 180
47. *In each example above, examine the relationship between the canonical representation of* lcm(m, n) *and those of* m *and* n.
48. *Describe a method for determining* lcm(m, n) *given the canonical representations of* m *and* n.

Definition The *greatest common divisor* of m and n, denoted gcd(m, n) is the largest number v that is a factor of both m and n.

49. *Determine the greatest common divisor of each of the following.*
    (i) 36 *and* 60
    (ii) 105 *and* 273
    (iii) 576 *and* 2016
50. *In each example in question 49, examine the relationship between the canonical representation of* gcd(m, n) *and those of* m *and* n.
51. *Describe a method for determining* gcd(m, n) *given the canonical representations of* m *and* n.
52. *Prove that* gcd(m, n) $\times$ lcm(m, n) $= mn$.

Note: If gcd(m, n) $= 1$, then m and n are said to be *relatively prime*.

1.5. Additional Questions

1. Prove that the square of an integer is either a multiple of 4 or one more than a multiple of 4. Is the converse of this statement true?

2.   Prove that the difference between the squares of any two odd integers is divisible by 8.

3.   Prove that the sum of five consecutive integers is divisible by 5. Is it always true that the sum of n consecutive integers is divisible by n? If not, hypothesize for which values of n it is true.

4.   Prove that if $a \equiv b \bmod m$ and m is a multiple of n, then $a \equiv b \bmod n$.

5.   In the Division Algorithm, we required that a and b be positive integers. Is the theorem still true if $a < 0$? if $b < 0$?

6.   Guess a formula for $\sum_{j=1}^{n} j(j!)$ and prove it by induction.

7.   Let f_n be the nth Fibonacci number. Hypothesize and prove formulas for each of the following.

$$\text{(a) } \sum_{j=1}^{n}(-1)^{j}f_{j} \qquad \text{(b) } \sum_{j=1}^{2n-1} f_{j}f_{j+1} \qquad \text{(c) } \sum_{j=1}^{n}f_{3j}$$

8.   Prove that if p is a prime greater than 3, then $p^2 - 1$ is divisible by 24.

9.   (a) Find the remainder when $1^{99} + 2^{99} + 3^{99} + 4^{99} + 5^{99}$ is divided by 5.
(b) Generalize the result of part (a). You need not prove your claim. (There are many possibilities.)

10.   If N is a perfect square, what must be true about the canonical representation of N?

11.   Find all primes p such that $p^{1996} + p^{1997}$ is a perfect square and prove that you have them all.

12.   Observe that $1^{2} = \dfrac{1(2)(3)}{6}$, $1^{2} + 3^{2} = \dfrac{3(4)(5)}{6}$, $1^{2} + 3^{2} + 5^{2} = \dfrac{5(6)(7)}{6}$.
Propose a general formula suggested by these examples and prove it.

13.   Prove that there are no integral values of n such that $5n^2 + 15$ is divisible by 17. [*Hint:* Let $n = 17k + r$, where $-8 \le r \le 8$. Consider each possible value of r. Does this account for all possibilities for n?]

14.   Consider the sequence $\{0, 1, 1, 2, 2, 3, 3, 4, 4, \ldots\}$ (in which each positive integer appears twice). Let $S(n) =$ the sum of the first n terms of this sequence.
(a) Derive a formula for $S(n)$.
(b) Show that $S(m + n) - S(m - n) = mn$ for all integers $m > n$.

15.   Prove that if p and $p + 2$ are both prime, then $p + 1$ is divisible by 6.

16.   Prove that N is a perfect square if and only if it has an odd number of divisors.

17. Find the two smallest numbers with exactly 8 divisors.

18. An integer is said to be *abundant* if $\sigma(n) > 2n$ and *deficient* if $\sigma(n) < 2n$, where $\sigma(n)$ is the sum of the divisors of n.
   (a) Determine the smallest positive abundant number.
   (b) Show that if $n = p^k$, where p is prime, then n is deficient.
   (c) Show that any multiple of an abundant number is abundant.

19. A number is said to be *superperfect* if $\sigma(\sigma(n)) = 2n$.
   (a) Show that 16 is superperfect.
   (b) Show that if $n = 2^q$, where $2^{q+1} - 1$ is prime, then n is superperfect.
   (c) Find the next superperfect number of this form.

20. A *triangular number* is a number of the form $\dfrac{n(n+1)}{2}$. Show that every even perfect number is triangular.

21. Show that if m and n are relatively prime, then $\tau(mn) = \tau(m)\tau(n)$, where $\tau(n)$ is the number of divisors of n.

22. Prove that the product of the divisors of an integer n is given by $n^{\tau(n)/2}$.

23. (a) Determine the remainder when 2^{38} is divided by 127. [*Hint:* $2^7 \equiv 1$ mod 127.]
   (b) Compute the last two digits of 3^{985}. [*Hint:* Work mod 100. Be persistent.]

24. (a) Let $\sum_{j=1}^{n} \dfrac{1}{j(j+1)}$. Use the fact that $\dfrac{1}{j(j+1)} = \dfrac{1}{j} - \dfrac{1}{j+1}$ to derive a formula for S_n.
   (b) Derive a formula for $\sum_{j=1}^{n} \dfrac{1}{j(j+2)}$ and prove it directly or by induction.

25. The Division Algorithm (Theorem 1.2) claims the existence of a quotient q and remainder r such that $a = bq + r$, for any positive integers a and b. In the text, we proved that q and r are unique. Here we prove that they exist.
   Let $S = \{a - xb \mid x \text{ is an integer and } a - xb \geq 0\}$.
   (a) Show that S is nonempty.
   (b) The Well-Ordering Principle says that any nonempty set of nonnegative integers has a smallest element. For the set S, let the smallest element be r. Argue that there exists q such that $r = a - qb$.
   (c) Show that $r < b$. [Hint: Assume the contrary and show that $r - b$ is an element of S. Where is the contradiction?]

Chapter 2
Numbers and Numerals

By now, you should have a fairly good idea of how the four basic processes—experimentation, conjecture, proof, and generalization—are used to create mathematics, at least in the context of integers and sequences of integers. In the remaining chapters, we'll apply the same concepts to other mathematical topics. In this chapter, we'll look at a larger set of numbers—the rational numbers, which contain the integers and many nonintegers (but not every real number). We'll ask what properties of the integers apply to the rationals and what other properties the rationals may have.

One interesting question concerns determining whether a number is rational. We'll give two answers: One involves the decimal representation of the number (and should be familiar), the other involves continued fractions (and is probably new to you).

In the last section of this chapter, we'll investigate the systems of numeration used to represent numbers.

2.1. Rational Numbers

In Chapter 1, we investigated the structure and properties of the integers. We saw that the integers are closed under addition, multiplication, and subtraction, but not under division. In other words, the quotient of two integers is a number that is not (necessarily) an integer. In this chapter, we'll investigate the set of numbers that arise from dividing two integers. We begin with a definition.

Definition x is said to be a *rational number* iff x can be expressed in the form $\frac{p}{q}$, where p and q are integers and $q \neq 0$.

An *irrational number* is one that is not rational. The set of rational numbers is usually denoted **Q** (for quotient).

The key phrase in the definition is "can be expressed." In some cases, it is obvious that a given number can be expressed in the desired form. For example, $0.75 = \frac{3}{4}$, $\sqrt{0.36} = 0.6 = \frac{3}{5}$ and $\sin\left(\frac{\pi}{6}\right) = 0.5 = \frac{1}{2}$ are clearly rational. On the other hand, it is less obvious whether repeating decimals such as $0.131313\ldots$ or numbers such as π, e, or $\sqrt{2}$ are rational or irrational.

With the integers, we had to assume various properties such as closure. Here, since the rationals are defined in terms of integers, we should be able to prove such properties.

Note: Henceforth, we'll assume that we can exclude division by 0 when trying to prove that a set is closed under division.

1. Prove that the rational numbers are closed under addition, multiplication, and division.

2. Are the integers rational?

Since the rationals are closed under division, then the rationals are *structurally different* than the integers. Concepts such as the Division Algorithm do not make sense because when we divide two rationals, we always get a unique, exact quotient and no remainder. Furthermore, every rational can be written as the product of other rationals in infinitely many ways; for example,

$$\frac{15}{14} = \left(\frac{3}{2}\right)\left(\frac{5}{7}\right) = \left(\frac{3}{7}\right)\left(\frac{5}{2}\right) = \left(\frac{2}{1}\right)\left(\frac{15}{28}\right) = \left(\frac{4}{3}\right)\left(\frac{45}{56}\right) = \cdots.$$

Thus, it is not clear what we would mean by primes and unique factorization.

The idea that different sets of numbers (or other mathematical objects) have different structures, and that there are questions or properties that pertain to one structure that don't pertain to others, is extremely important. Indeed, a large part of advanced mathematics consists of determining how various properties can be adapted to sets of different structures.

Now let's return to the question of determining whether a given number is rational or irrational. As we stated earlier, one answer involves the decimal representation of the number.

3. Pick several rational numbers (i.e., fractions) and convert them to decimal representation.

You should have gathered evidence to support the following theorem.

Theorem 2.1 x is a rational number if and only if x can be expressed as a decimal numeral that either terminates in a finite number of digits or contains a sequence of digits that repeats consecutively infinitely often.

We'll use the phrase "repeating decimal" to describe the case in which the decimal has a sequence of digits that repeats consecutively infinitely often. Note that this does not preclude the possibility of there being some digits at the beginning that don't repeat. For example, 0.253181818. . . is a repeating decimal and, if the theorem is true, must represent a rational number. Those repeating decimals that have some initial nonrepeating digits are said to be *delayed*. The word *consecutively* is important since it precludes the possibility that there are one or more digits that appear infinitely often but with other digits in between. So, if the theorem is true, the number

$$0.\underline{24}1\underline{24}3\underline{24}7\underline{24}8\underline{524}. . .$$

would not be rational.

Since Theorem 2.1 is an if-and-only-if statement, we have to prove the following two parts.

- If the decimal representation of x terminates or repeats, then x is rational.
- If x is rational, then its decimal representation terminates or repeats.

The proof of the first part is relatively straightforward; the second part requires a bit more machinery, which we will develop. In the process of proving the theorem, we'll also determine which rational numbers have repeating decimals and which of those are delayed. We'll also be able to determine the length of the repeating pattern and how long it is delayed, if at all.

4. *Prove that if the decimal representation of x terminates, then x is rational.*

Now assume that x has a nondelayed repeating decimal.

5. *Show that x can be expressed as an infinite geometric series. What is the ratio between consecutive terms?*

6. *Show that x is rational.*

7. *Modify the proof to account for delayed repeating decimals.*

8. *Express each of the following as fractions.*
   (i) $0.\overline{35}$ *(ii)* $0.\overline{271}$ *(iii)* $0.1\overline{4}$

We've now completed the proof of the "if" part of Theorem 2.1.

To prove the "only if" part, let's begin by characterizing those rationals whose decimals terminate.

9. *What must be true about the denominator of a rational number whose decimal terminates?*

10. Complete and prove the following theorem.

Theorem 2.2   The rational number $\frac{p}{q}$, reduced to lowest terms, has a terminating decimal representation iff $q = $ _____. The number of digits in the decimal representation is _____.

It remains to show that those rational numbers with denominators not satisfying Theorem 2.2 always have repeating decimal representations. We'll also need to distinguish between delayed and nondelayed repeating decimals.

Here is an instance where numerical evidence may not be entirely convincing, especially if it is subject to the limitations of a typical calculator. For example, in

$$\frac{11}{19} = 0.\overline{578947368421052631},$$

there is an 18-digit repeating pattern that won't be apparent in a standard 8-digit display. Worse yet, there is a 46-digit repeating pattern in

$$\frac{19}{47} = 0.\overline{4042553191489361702127659574468085106382297872}.$$

11. What are the possible denominators of rational numbers with a one-digit nondelayed repeating decimal?

12. What are the possible denominators of rational numbers with a two-digit nondelayed repeating decimal?

13. More generally, what are the possible denominators of rational numbers with an n-digit nondelayed repeating decimal?

The proof of the statement in question 13 relies on the fact that a decimal with an n-digit repeating pattern is really a geometric series with common ratio $r = 10^{-n}$. As we learned in Chapter 1, since $|r| < 1$, this series converges and its sum is

$$\frac{a}{1-r} = \frac{a}{1-10^{-n}} = \frac{a \cdot 10^n}{10^n - 1}.$$

Depending on the value of a, this may or may not be reduced to lowest terms. In either case, we have the following theorem.

Theorem 2.3   If $\gcd(q, 10) = 1$ and $\gcd(p, q) = 1$, then the rational number $\frac{p}{q}$ has a nondelayed repeating decimal expansion with n digits, where n is the smallest positive integer such that $10^n - 1$ is divisible by q.

The remaining question is: How do we know that such an n always exists? In other words, for every q, relatively prime to 10, is there an n such that $10^n - 1$ is divisible by q?

Let's look at some numerical evidence. Numbers of the form $10^n - 1$ have decimal numerals consisting of n 9s. The table below has the canonical representations of these numbers for $n = 2, 3, \ldots, 12$.

2	$3^2 \times 11$
3	$3^3 \times 37$
4	$3^2 \times 11 \times 101$
5	$3^2 \times 41 \times 271$
6	$3^3 \times 7 \times 11 \times 13 \times 37$
7	$3^2 \times 239 \times 4649$
8	$3^2 \times 11 \times 73 \times 101 \times 137$
9	$3^4 \times 37 \times 333{,}667$
10	$3^2 \times 11 \times 41 \times 271 \times 9091$
11	$3^2 \times 21{,}649 \times 513{,}239$
12	$3^3 \times 7 \times 11 \times 13 \times 101 \times 9901$

Although there is a wide variety of primes included in the canonical representations, there are some relatively small primes—such as 17, 19, and 23—that do not appear among the first 12 values of n. We need to prove that these primes are guaranteed to show up for some n if we continue the table further.

To complete the proof, we need to introduce a theorem that can be found in most elementary number theory texts but whose proof is beyond the scope of this text. First, a definition.

Definition   Let $\varphi(m)$ = number of integers less than or equal to m that are relatively prime to m.

For example, $\varphi(8) = 4$ since there are four integers—1, 3, 5, and 7—that are less than or equal to and relatively prime to 8.

14. What is $\varphi(10)$? $\varphi(18)$?

15. What is $\varphi(p)$, where p is prime? What is $\varphi(p^k)$, where p is prime and k is an integer?

The function $\varphi(m)$ is called *Euler's Phi-Function* (or sometimes *Euler's totient function*). It has many interesting properties, one of which is in the following theorem.

Theorem 2.4 (Euler-Fermat Theorem)   If a and m are relatively prime, then $a^{\varphi(m)} \equiv 1 \bmod m$.

Although a general proof is beyond the scope of this text, we can prove Theorem 2.4 for specific values of m.

16. Prove Theorem 2.4 for $m = 3$ and $m = 5$.

Now let $a = 10$ and $m = q =$ the denominator of the given rational number. Theorem 2.4 says that if q is relatively prime to 10 (meaning that the canonical representation of q contains no factors of 2 or 5), then there exists an integer $n = \varphi(q)$ such that $10^n - 1$ is divisible by q. But this is precisely what we needed to show to prove Theorem 2.3.

Consequently, any rational number with denominator q that is relatively prime to 10 has a nondelayed repeating decimal representation with at most $\varphi(q)$ digits. Moreover, it can be shown that if the number of repeating digits is strictly less than $\varphi(q)$, then it must be a divisor of $\varphi(q)$. For example, $\frac{7}{13} = 0.\overline{538461}$ has a repeating pattern of length 6 and 6 is a divisor of $\varphi(13) = 12$.

17. Is this result consistent with the 18-digit and 46-digit repeating pattern examples given after Theorem 2.2?

18. What is the longest possible repeating pattern for the rational number $\frac{p}{q}$?

We've now shown that rational numbers with denominators whose only factors are 2 and 5 have terminating decimals, and those with denominators with no factors of 2 and 5 have nondelayed repeating decimals. The only other case to consider is those rational numbers whose denominators have at least one factor of 2 or 5 and some prime other than 2 and 5 as factors. These must have delayed repeating decimals.

Let's look at an example such as $\frac{1}{6}$. By multiplying numerator and denominator by 5 and factoring, we get $\frac{1}{6} = \frac{1}{10}(\frac{5}{3})$. The fraction $\frac{5}{3}$ has a denominator that is relatively prime to 10; hence, its decimal representation consists of an integer followed by a nondelayed repeating decimal, $1.666.\ldots$ Upon multiplying by $\frac{1}{10}$, we get the delayed repeating decimal $0.16666.\ldots$

This argument can be generalized. Suppose $\frac{p}{q}$ is a rational number such that $q = 2^a 5^b r$, where r is relatively prime to 10. Then

$$\frac{p}{q} = \frac{p}{2^a 5^b r} = \frac{1}{10^d}\left(\frac{2^{d-a}5^{d-b}p}{r}\right),$$

where $d = \max(a, b)$—that is, $d = a$ if $a \geq b$ and $d = b$ if $a \leq b$.

19. Complete the proof.

20. How many digits will there be before the pattern starts repeating?

2.2. Irrational Numbers

As we said earlier, an *irrational number* is, by definition, one that is not rational. In other words, an irrational number cannot be expressed as the ratio of two integers. In view of the results of Section 2.1, we can say that an irrational number has a decimal representation that neither terminates nor repeats.

Proving that a given number is irrational is not always an easy task. (In fact, proving something can't be done in mathematics is often more difficult than proving that it can.) With modern computers, we can determine a large number of digits in the decimal representation, but we cannot tell whether the representation might eventually repeat. Indeed, the repeating pattern could contain thousands of digits, or it could be delayed by a few million digits.

Nonetheless, proving irrationality in some cases is relatively simple. For example, consider $\sqrt{2}$. To prove that $\sqrt{2}$ is irrational, we proceed by contradiction. Assume that $\sqrt{2}$ is rational; that is, there exists relatively prime integers p and q such that $\sqrt{2} = \frac{p}{q}$ or, equivalently, $p^2 = 2q^2$.

1. Why can we assume that p and q are relatively prime?

2. Argue that p must be even.

3. Show that if p is even, then q must be even.

4. Where is the contradiction?

We can use this technique to prove the irrationality of other square roots, cube roots, and so on. It is more difficult to show that numbers such as π or e are irrational.

Now let's investigate the structure of the irrational numbers.

5. Are the irrationals closed under addition? If so, prove it. If not, provide a counterexample. What about multiplication?

Now suppose x is an irrational number and r is a rational number. Then $x + r$ is irrational. To see why, suppose $x + r$ were a rational number s. Then $x = s - r$.

6. Where is the contradiction?

7. Prove that $\frac{x}{r}$ and $\frac{r}{x}$ are both irrational.

8. Show that $\dfrac{\sqrt{5} + 3}{7}$ is irrational.

2.2.1. Extensions of the Rationals

The entire set of irrationals is of limited interest. It has very little structure (unlike the integers and rationals) and, other than proving certain numbers are irrational, there aren't many interesting questions we can answer at this level.

However, there are sets that are "mixtures" of rationals and irrationals that have a much richer structure, leading to some more fruitful investigations. For example, let

$$S = \{a + b\sqrt{2}, \text{ where } a \text{ and } b \text{ are rational}\}.$$

S is said to be an *extension* of the rationals.

  9. *Are the rational numbers elements of S?*

  10. *Decide whether S is closed under each of the following operations. If so, prove it; if not, provide a counterexample.*
      (i) *addition*
      (ii) *multiplication*
      (iii) *division*

  11. *Under what conditions is the product of two elements of S rational?*

It should be clear that, as far as the closure properties are concerned, there is nothing special about $\sqrt{2}$; indeed, we could have extended the rationals with $\sqrt{k}$ for any $k > 0$ that is not a perfect square. A natural question to ask is what would happen if we extended the rationals with an irrational cube root, say, $\sqrt[3]{2}$. So, let

$$S = \{a + b\sqrt[3]{2}, \text{ where } a \text{ and } b \text{ are rational}\}.$$

  12. *Show that S is closed under addition but not under multiplication.*

  13. *Determine a value of x such that the set*

$$S' = \{a + b\sqrt[3]{2} + cx, \text{ where } a, b, \text{ and } c \text{ are rational}\}$$

is closed under multiplication. Prove your claim.

Rather than extending the rationals, we could extend the integers with an irrational number. Let

$$S = \{a + b\sqrt{2}, \text{ where } a \text{ and } b \text{ are integers}\}.$$

Clearly, S is closed under addition and multiplication.

  14. *Show that S is not closed under division.*

Here is a set that has the same structure as the integers. So it makes sense to ask whether other properties such as the Division Algorithm and unique factorization apply. These are nontrivial questions that we won't be able to completely answer at this point. To see why they are difficult questions, observe that $14 = 2 \times 7 = (4 + \sqrt{2})(4 - \sqrt{2})$. At first glance, we might think that this contradicts unique factorization. However, this conclusion is premature since we have not determined what we mean by "prime" in this context. It is not as simple as saying that primes are those elements that cannot be written as the product of two other elements of S.

In fact, neither factorization of 14 shown above is the product of "primes" since $2 = (2 + \sqrt{2})(2 - \sqrt{2})$, $7 = (3 + \sqrt{2})(3 - \sqrt{2})$, $4 + \sqrt{2} = (2 + \sqrt{2})(3 - \sqrt{2})$, and $4 - \sqrt{2} = (3 + \sqrt{2})(2 - \sqrt{2})$. Notice, in particular, that integers that are prime in the usual sense, such as 2 and 7, are not necessarily prime in S. It turns out that S does indeed have unique factorization. It takes a bit of work to show that $3 + \sqrt{2}$, $3 - \sqrt{2}$, $2 + \sqrt{2}$, and $2 - \sqrt{2}$ are all prime in S and, hence, the unique factorization of 14 is $14 = (3 + \sqrt{2})(3 - \sqrt{2}) \times (2 + \sqrt{2})(2 - \sqrt{2})$. We'll investigate this in more detail in Chapter 7.

2.3. Continued Fractions

We have characterized rational numbers as those numbers with terminating or repeating decimal representations, and irrational numbers as "everything else" (among the real numbers). In this section, we'll look at an alternative characterization in terms of *continued fractions*.

A *simple continued fraction* is an expression of the form

$$a_0 + \cfrac{1}{a_1 + \cfrac{1}{a_2 + \cdots}},$$

where $a_0, a_1, \ldots$ are positive integers. The continued fraction may terminate, in which case it is said to be *finite*, or it may continue indefinitely, in which case it is *infinite*. To save space, we'll use the notation $[a_0, a_1, \ldots]$ to represent the continued fraction.

It is possible that the last entry in the continued fraction could be a noninteger. So, for example, we might write $[1, 3, 5, \frac{7}{3}]$ or $[2, 3, \pi]$. Such continued fractions are not "simple".

2.3.1. Finite Continued Fractions

If the continued fraction is finite, then it is easy to evaluate by performing the indicated arithmetic, starting from the bottom and working upward.

1. What rational number is represented by each of the following continued fractions?

(i) [1, 2, 3]   *(ii)* [3, 2, 4, 1]   *(iii)* [1, 1, 1, 1, 1]

2. Show that [1, 2, 3] = [1, 2, 2, 1].

3. The result in question 2 is an example of a more general result. State and prove this generalization.

Conversely, given a rational number, we can find its continued fraction representation.

4. Express each of the following rational numbers as a simple finite continued fraction.

(i) $\frac{7}{5}$ *(ii)* $\frac{11}{7}$ *(iii)* $\frac{45}{14}$ *(iv)* 3.25

At this point, we may be led to believe the following theorem.

Theorem 2.5 x is rational if and only if any continued fraction representation of x is finite.

The proof of the "if" part is trivial since evaluating a finite continued fraction consists of a finite sequence of additions and divisions of rational numbers. By the closure properties of the rationals, this must lead to a rational number.

For the "only if" part, it is certainly clear that every rational (and, indeed, every real number) can be expressed as a continued fraction. We need to prove that the continued fraction is finite; in other words, we need to show that the process you used to answer question 4 always terminates.

Suppose $x = \frac{p}{q}$, where p and q are positive integers. By the Division Algorithm, there exist unique positive integers a_0 and r_0, where $0 \le r_0 < q$, such that $p = a_0 q + r_0$.

5. What happens if $r_0 = 0$?

If $r_0 \ne 0$, then we can apply the Division Algorithm to the quotient of q and r_0; that is, there exist integers a_1 and r_1, where $0 \le r_1 < r_0$, such that $q = a_1 r_0 + r_1$. If $r_1 \ne 0$, then $r_0 = a_2 r_1 + r_2$, where $0 \le r_2 < r_1$.

This process of applying the Division Algorithm to the quotient of the two previous remainders must terminate in a finite number of steps. In other words, there exists some k such that $r_k = 0$.

6. Why?

7. Show that $x = [a_0, a_1, \ldots, a_k]$.

This completes the proof of Theorem 2.5. Note that since the Division Algorithm ensures a unique quotient and remainder, then we have shown that the continued fraction representation of any rational number is "essentially unique." The exception to uniqueness is the result described in question 3.

Now let $x = [a_0, a_1, \ldots, a_n]$ be a finite continued fraction. If n is large, it may be tedious to calculate the value of the continued fraction. Thus, it may be of interest to see what happens if we "chop off" the fraction after k terms, where $k < n$. This motivates the following definition.

Definition The kth *convergent* to the continued fraction $x = [a_0, a_1, \ldots, a_n]$ is $c_k = [a_0, a_1, \ldots, a_k]$.

Since the convergents are all finite continued fractions themselves, they must be rational and we shall write

$$c_k = \frac{p_k}{q_k}.$$

So that p_k and q_k are themselves well defined, we shall agree to express c_k in lowest terms. The zeroth convergent is $c_0 = a_0$; hence, $p_0 = a_0$ and $q_0 = 1$. Of course, the nth convergent is just x itself.

8. *Compute the convergents for each of the following. Note how the values of the successive convergents relate to the each other and to the value of the entire continued fraction.*

(i) [1, 2, 3, 1]   *(ii)* [3, 4, 1, 2, 5]   *(iii)* [1, 1, 1, 1, 1]

9. *(i) Derive an expression for p_1 in terms of p_0 and a_1. Derive a similar expression for q_1 in terms of q_0 and a_1.*
(ii) Derive an expression for p_2 in terms of p_0, p_1, and a_2. Derive a similar expression for q_2 in terms of q_0, q_1, and a_2.

10. *The relationship above can be generalized to give recursive expressions for p_k in terms of p_{k-1}, p_{k-2}, and a_k, and for q_k in terms of q_{k-1}, q_{k-2}, and a_k. Write those in the following theorem.*

Theorem 2.6 For $k \geq 2$, the kth convergent to the continued fraction $x = [a_0, a_1, \ldots, a_n]$ satisfies the relationship $p_k =$ _____ and $q_k =$ _____.

Theorem 2.6 can be extended to account for the cases $k = 0$ and 1, by appropriately defining p_{-1}, q_{-1}, p_{-2}, and q_{-2}.

11. *Do so.*

12. *Use Theorem 2.6 to compute the value of* [2, 1, 4, 7, 6].

The proof of Theorem 2.6 is by induction. We've already shown that it is true for $k = 2$. To complete the proof, we need to use the fact that

$$[a_0, a_1, \ldots, a_k, a_{k+1}] = [a_0, a_1, \ldots, b_k], \quad \text{where } b_k = a_k + \frac{1}{a_{k+1}}.$$

So, the $(k + 1)$st convergent to $[a_0, a_1, \ldots, a_k, a_{k+1}]$ is equal to the kth convergent to $[a_0, a_1, \ldots, b_k]$. Note that $[a_0, a_1, \ldots, b_k]$ is a nonsimple continued fraction since the last entry is not necessarily an integer; this does not, however, affect the validity of Theorem 2.6.

13. Finish the induction part of the proof.

To finish the proof, we need to show that for all k, p_k, and q_k—when computed by Theorem 2.6—are relatively prime. To do so, we first need a corollary.

14. Complete and prove by induction the following corollary.

Corollary 2.1 $p_k q_{k-1} - p_{k-1} q_k =$ ——————.

To prove that p_k and q_k are relatively prime, suppose the contrary—that is, there exists some integer $d > 1$ such that p_k and q_k are divisible by d.

15. Show that this leads to a contradiction to Corollary 2.1.

The next corollary follows immediately from Corollary 2.1.

Corollary 2.2 $c_k - c_{k-1} = \dfrac{(-1)^{k-1}}{q_k q_{k-1}}.$

16. Use Theorem 2.6 and Corollary 2.1 to prove the following corollary.

Corollary 2.3 $c_k - c_{k-2} = \dfrac{(-1)^k a_k}{q_k q_{k-2}}.$

17. What does Corollary 2.3 say about the sequence of even-numbered convergents? odd-numbered convergents?

18. Combining Corollaries 2.2 and 2.3, describe the behavior of the entire sequence of convergents. Which is the largest? the smallest?

19. Describe how to use the first few convergents to approximate the value of a continued fraction.

There is one more corollary we need to make the discussion complete. The proof follows directly from Theorem 2.6 and the fact that $a_k \geq 1$ for all k.

Corollary 2.4 $q_0 \leq q_1 < q_2 < q_3 < \cdots$

20. Complete the proof.

21. Argue that $q_k \geq k$ for all k.

2.3.2. Infinite Continued Fractions

Now let's consider infinite continued fractions of the form $[a_0, a_1, a_2, \ldots]$. The first question we need to answer is how to interpret such an expression. Finite continued fractions are evaluated "from the bottom up"; here, there is no "bottom."

The situation is not unlike the one encountered when defining infinite series of the form $\Sigma_{j=1}^{\infty} u_j$. We saw an example of this in Chapter 1 when we studied infinite geometric series. The key question is whether the series converges. To answer this question, we investigate the behavior of the sequence of partial sums of the form $s_n = \Sigma_{j=1}^{n} u_j$. If $\lim_{n\to\infty} s_n$ exists, then the infinite series converges to the value of that limit. If $\lim_{n\to\infty} s_n$ does not exist, then the infinite series diverges. For the case of geometric series, the partial sums are

$$s_n = \sum_{j=1}^{n} ar^{j-1} = \frac{a(1 - r^n)}{1 - r}.$$

If $|r| < 1$, then $\lim_{n\to\infty} s_n$ exists and hence the infinite series converges to $\frac{a}{1 - r}$.

22. What quantity for infinite continued fractions is analogous to the partial sums for infinite series? Write a definition of an infinite continued fraction in terms of limits.

Next we need to determine the conditions under which infinite continued fractions converge. Theorem 2.6 and its corollaries tell us that

- the odd-numbered convergents form a strictly decreasing sequence.
- the even-numbered convergents form a strictly increasing sequence.
- every even-numbered convergent is less than every odd-numbered convergent.

Thus, all the odd-numbered convergents are bigger than the value of the continued fraction and all the even-numbered convergents are smaller than the value of the continued fraction. These facts ensure that the sequence of odd-numbered convergents and the sequence of even-numbered convergents both converge—that is, $\lim_{k\to\infty} c_{2k+1}$ and $\lim_{k\to\infty} c_{2k}$ both exist.

Next, we'd like to show that these two limits are the same. To do so, we'll show that the absolute value of the difference between the limits must be 0.

$$\left| \lim_{k \to \infty} c_{2k+1} - \lim_{k \to \infty} c_{2k} \right| = \lim_{k \to \infty} \left| c_{2k+1} - c_{2k} \right| = \lim_{k \to \infty} \left| \frac{1}{q_{2k} q_{2k+1}} \right|,$$

where the second equality comes from Corollary 2.2.

23. *Continue the proof to show* $\left| \lim_{k \to \infty} c_{2k+1} - \lim_{k \to \infty} c_{2k} \right| = 0.$

We have now shown that infinite continued fractions *always converge*. Moreover, in view of Theorem 2.5, they must converge to an irrational number.

24. *Write the first five terms of the infinite continued fraction expansion of each of the following.*

  (i) π *(ii)* e *(iii)* $\sqrt{2}$

It is difficult to deal with infinite continued fractions in general. However, a special case—those in which some set of terms repeats consecutively infinitely often—is easier to deal with. Such fractions are said to be *periodic*. An example of a periodic continued fraction is $[5, 4, \overline{1, 2}] = [5, 4, 1, 2, 1, 2, 1, 2, \ldots]$.

Consider the example

$$[\overline{1, 2}] = 1 + \cfrac{1}{2 + \cfrac{1}{1 + \cfrac{1}{2 + \cdots}}}.$$

Let x represent the value of the continued fraction. Note that the boldface part is also equal to x. Hence,

$$x = 1 + \cfrac{1}{2 + \cfrac{1}{x}}.$$

25. *Solve this for* x.

26. *Determine the value of each of the following periodic continued fractions.*

  (i) $[\overline{1}]$ *(ii)* $[\overline{1, 3, 2}]$ *(iii)* $[1, \overline{3, 2}]$

All your answers to questions 25 and 26 should be of the form $\dfrac{a + b\sqrt{c}}{d}$,

where $a, b,$ and d are integers and c is a positive integer that is not a perfect square. Numbers of this form are called *quadratic surds*. We propose the following:

Theorem 2.7 x is a quadratic surd iff x can be represented as a periodic continued fraction.

We shall prove the "if" part of this fact in the special case that the continued fraction is "purely periodic"; that is, it contains no nonrepeating part at the beginning.

27. Let $x = [\overline{a_0, a_1, \ldots, a_k}]$. Argue that $x = [a_0, a_1, \ldots, a_k, x]$, where the expression on the right is a nonsimple finite continued fraction.

28. Argue that $[a_0, a_1, \ldots, a_k, x]$, when simplified, is the ratio of two linear functions of x—that is, of the form $\dfrac{\alpha x + \beta}{\gamma x + \delta}$. Furthermore, show that $\alpha = p_k$, $\beta = p_{k-1}$, $\gamma = q_k$, and $\delta = q_{k-1}$, where $\dfrac{p_k}{q_k}$ and $\dfrac{p_{k-1}}{q_{k-1}}$ are the last two convergents to $[a_0, a_1, \ldots, a_k]$.

29. Show that x is a quadratic surd.

The proof can be modified to account for the case in which the repeating pattern is delayed. We omit the details here. We also omit the "only if" part of Theorem 2.7 since it is rather messy.

30. Characterize all quadratic surds whose continued fraction representations consist of a single periodic entry.

31. Find the continued fraction representation of $\sqrt{2}$ and prove that it is correct.

It is also worth noting that the convergents to the continued fraction representation of an irrational number are the "best" rational approximations to that number. What we mean is that if $c_n = \dfrac{p_n}{q_n}$ is the nth convergent to the continued fraction representation of x, then there are no rational numbers with denominators less than or equal to q_n that are closer to x in absolute value.

For example, the continued fraction representation of π is $[3, 7, 15, 1, 292, \ldots]$, so the first few convergents are $c_1 = \frac{22}{7} = 3.142857$, $c_2 = \frac{333}{106} = 3.141509$, and $c_3 = \frac{355}{113} = 3.1415929$, which is correct to 6 decimal places. Moreover, there are no rational numbers with denominators less than or equal to 113 that are closer to π.

2.4. Systems of Numeration

There are many terms in mathematics that cannot be defined in the usual sense. Some examples are *point*, *line*, and *plane*. Although we can't define these terms, we can represent them in a variety of ways. In the examples above, for instance, we can draw a picture.

Another undefinable concept is *number*. We've used the term many times so far, but at no time did we write a definition of it. Still, we can represent numbers in several ways. We can use words for them, such as *one, two, three*, and so on. We can draw pictures for them, such as

to represent the number *three*. Obviously, using pictures or words becomes inconvenient for large numbers. So we use *numerals* instead. Numerals are symbols that represent numbers. There have been many systems of numeration used throughout history. The Romans, Babylonians, and Egyptians, among others, each had their own systems.

Nowadays we use numeration systems that are *positional*, meaning that the value of the symbols used depends on their location—or position—within the numeral. (In contrast, the Roman system was not positional. An X in Roman numerals represents 10 no matter where it appears in the numeral, although sometimes it is added to what follows—as in $XV = 5 + 10 = 15$—and sometimes it is subtracted—as in $XC = 100 - 10 = 90$.)

To create a positional numeration system, we first have to specify the *base* of the system. We generally express numbers as base 10 (decimal) numerals. We use base 10 mainly because most humans have 10 fingers. Thus, when we write a numeral such as 372 in base 10, the "2" means 2×10^0, the "7" means 7×10^1, and the "3" means 3×10^2.

1. In general, what is the value of a digit d that appears r places to the left of the decimal point?

2. In general, what is the value of a digit d that appears r places to the right of the decimal point?

3. What different digits do we use in base 10?

Suppose we decide to express numbers as numerals in a different base; say base *b*, where *b* is a positive integer not equal to 1.

4. What different digits do we need to use?

5. What is the value of a digit d that appears r places to the left of the "decimal" point? r places to the right?

6. Express each of the following as a base 10 numeral (the subscript indicates the base).

(i) 314_9 *(ii)* 1001101_2 *(iii)* 210.12_3 *(iv)* $12.\overline{1}_4$

7. For what base b is $271_b = 226_{10}$?

8. *For what base b is* $(24_b)^2 = 554_b$?

9. *Show that* 10201_b *is a perfect square in every base* $b \geq 3$. *What is its square root?*

Be careful how you read numerals. For example, the numeral in 6(*i*) is properly read "three-one-four, base 9," not "three hundred fourteen."

For bases greater than 10, we need additional symbols to represent the digits. Typically we use letters of the alphabet. So, for example, in base 16, the digits are

$$\{0, 1, 2, 3, 4, 5, 6, 7, 8, 9, A, B, C, D, E, F\},$$

where A represents ten, B represents eleven, and so on.[1]

10. *What number does* $B2F_{16}$ *represent?*

Now let's develop an algorithm for determining the base b representation of a number N.

11. *According to the Division Algorithm, we can write* $N = q_0 b + r_0$, *where* $0 \leq r_0 \leq b - 1$. *Argue that* r_0 *is the first digit (on the right) of the base b numeral for N.*

12. *Now look at the first quotient* q_0. *Again by the Division Algorithm, we can write* $q_0 = q_1 b + r_1$. *Argue that* r_1 *is the second digit (from the right) in the base b numeral for N.*

13. *Describe how to get the remaining digits. When can we stop?*

14. *Use this algorithm to express the decimal numeral* 259 *in base 3 and base 6.*

15. *This algorithm only works for integers. Why?*

One of the advantages to using positional notation is that it makes doing arithmetic much easier. (If you don't believe this, try multiplying CCLXII by XXXIV without first converting to a positional system.)

---

[1] At one time, there was a society called The Duodecimal Society of America who advocated the use of base 12. Part of their rationale was that we already had some vestiges of it, such as 12 inches in a foot and the use of the words *dozen* and *gross* to represent 12 and 12^2. They published a small journal called *The Duodecimal Bulletin*. Among the symbols suggested to represent ten and eleven were * (which looks something like the letter X, the Roman numeral for ten) and # (which looks like two perpendicular 11s). Curiously, those symbols appear on the standard Touch-Tone telephone.

Here are a few exercises so you can practice doing arithmetic in other bases.[2] Try to do the arithmetic directly in the given base, although you can check your answers by converting to base 10.

16. (i) $123_5 + 234_5$ (ii) $110011_2 \cdot 11_2$
     (iii) $2032_8 \div 6_8$ (iv) $\sqrt{64}_{16}$

2.4.1. Rational Numbers in Other Bases

It is important to realize that the terms *rational* and *irrational* used earlier in this chapter apply to numbers and are unaffected by the system of numeration chosen to express the numbers. Nonetheless, one criterion for determining whether a number is rational is to examine its decimal numeral representation. Specifically, x is rational iff its decimal representation terminates or repeats. The question is whether that is true for other bases.

It should be clear that the proof of Theorem 2.1 can be modified to other bases. What differs is the distinction between those rational numbers with base b representations that terminate and with those that repeat. For example, $\frac{1}{3} = 0.333\ldots_{10} = 0.1_3 = 0.2_6$. So, some rational numbers with repeating decimal representations have terminating representations in other bases, and vice versa.

17. In which bases does $\frac{1}{3}$ have a terminating representation?

18. Which rational numbers terminate in base 2? in base 3? in base 6?

19. Which rational numbers terminate in base b? Which have nondelayed repeating representations? Which have delayed repeating representations?

20. In those rationals with repeating representations, how long is the repeating pattern?

The algorithm derived earlier for converting integers from base 10 to base b will not work for nonintegers since it relies on the Division Algorithm, a property that does not apply to nonintegers. Let's develop an algorithm that does work for nonintegers.

Let x be the number we wish to convert to base b. We can assume that $0 < x < 1$ since, if x were greater than 1, its "integer part" could be converted by the previous algorithm. The "fractional part" has to be treated separately. Converting x to base b means that we want to find integers $a_1, a_2, \ldots$, with $0 \le a_i \le b - 1$ for all i, such that

---

[2] In his song "New Math," singer-songwriter Tom Lehrer, tries to teach the audience how to do arithmetic in base 8. He says, "Don't worry. Base 8 is just like base 10 . . . if you're missing two fingers." Please don't resort to self-mutilation to do these problems.

$$x = \frac{a_1}{b} + \frac{a_2}{b^2} + \frac{a_3}{b^3} + \cdots.$$

21. Argue that a_1 is the integer part of bx.

22. How would we get a_2? a_3?

23. Use this algorithm to convert each of the following.

(i) $\dfrac{3}{4}$ *to base 3*   *(ii)* $\dfrac{3}{4}$ *to base 5*   *(iii) 0.45 to base 8*

24. Use your answer to question 20 to determine how long the repeating pattern should be when $\frac{17}{35}$ is written in base 6. Verify.

2.5. Additional Questions

Questions $1 - 5$ pertain to the function $\varphi(n)$ = number of integers less than n that are relatively prime to n. (See Theorem 2.4.) You will need to make use of the fact that if m and n are relatively prime, then $\varphi(mn) = \varphi(m)\varphi(n)$.

1. Prove that if p is prime, then $\varphi(p^k) = p^k - p^{k-1}$.

2. (a) Use the result of question 1 to derive a formula for $\varphi(n)$, where $n = p_1^{a_1} p_2^{a_2} \cdots p_r^{a_r}$.
(b) Determine $\varphi(180)$, $\varphi(3003)$, and $\varphi(25,000)$.

3. (a) If n is an odd integer, how are $\varphi(2n)$ and $\varphi(n)$ related?
(b) If n is an even integer, how are $\varphi(2n)$ and $\varphi(n)$ related?

4. How are $\varphi(n)$ and $\varphi(n^2)$ related?

5. Prove that if $n > 2$, then $\varphi(n)$ is even. [*Hint:* If n is a power of 2, the result follows from question 1. If n is not a power of 2, then $n = p^k m$ where p is a prime not equal to 2 and p is relatively prime to m. Then use the result of question 2.]

6. Show that

$$\frac{1}{n} + \frac{1}{n+1} + \frac{1}{n+2}$$

is a nonterminating decimal for every positive integer n.

7. Let $S = \{a + b\sqrt{2} + c\sqrt{5}$, where a, b, and c are integers$\}$. Is S closed under multiplication? If not, "fix it" so that it is closed.

8. Let $S = \{a + b\sqrt[3]{2} + c\sqrt[3]{4}$, where a, b, and c are rational$\}$. Prove that for every nonzero element x in S, there exists a nonzero element y in S such that xy

is rational. (In fact, there are infinitely many possibilities for y.) Use this to prove that S is closed under division.

9. For the purposes of this problem, assume that we have already shown that $\sqrt{k}$ is irrational for every positive integer k that is not a perfect square. Our goal is to prove that $\sqrt{2} + \sqrt{3}$ is irrational.

(a) What is wrong with the following argument? "Since $\sqrt{2}$ and $\sqrt{3}$ are irrational, then $\sqrt{2} + \sqrt{3}$ is irrational."

(b) For a correct approach, assume $r = \sqrt{2} + \sqrt{3}$ is rational. Show that this implies $\sqrt{6} = \dfrac{r^2 - 5}{2}$. Why is this a contradiction?

10. Let $x = \log(2)$, where the logarithm is taken in base 10; that is, x is the real number such that $10^x = 2$. (The approximate value of x is 0.3010.) Show that x is irrational. [*Hint:* Assume $x = \frac{p}{q}$ is rational. Use the Fundamental Theorem of Arithmetic to show a contradiction.]

11. It should be clear that $[a_0 + 1, a_1, \ldots, a_k] = [a_0, a_1, \ldots, a_k] + 1$. How is $[a_0, a_1 + 1, \ldots, a_k]$ related to $[a_0, a_1 \ldots, a_k]$?

12. Express $[a_0, a_1] + [b_0, b_1]$ as a nonsimple continued fraction.

13. The first three entries in a continued fraction are $[3, 2, 1, \ldots]$. Is it possible that the value of the continued fraction is 3.6? Explain.

14. Show that the convergents to the continued fraction $[1, 1, 1, \ldots]$ are the ratios of successive Fibonacci numbers.

15. The equation used in question 25 of Section 2.3 to solve for x has two roots. Argue that only one of these roots is positive.

16. Consider the "continued radical"

$$\sqrt{2 + \sqrt{2 + \sqrt{2 + \sqrt{2 + \cdots}}}}.$$

We can approximate the value of this expression by computing a finite number of terms, much the way we approximate an infinite continued fraction by computing its convergents.

(a) Compute the first three approximations.

(b) Let u_n be the approximation obtained by using n radicals. Write a recursive equation that u_n must satisfy.

(c) Assuming the successive values of u_n converge, determine the value of the infinite continued radical. Does this agree with your approximations obtained in part (a)?

17. The first fifteen positive integers are multiplied together. How many zeroes are there at the end of the product when expressed in base 12?

18. If $47_a = 74_b$, find the smallest possible value of $a + b$.

19. Develop an algorithm for converting a base b numeral to a numeral in base b^2 without first converting to base 10. Prove that the algorithm works. Use your algorithm to express 1001100111_2 in base 4.

20. Let $S_b(n)$ be the sum of the digits in the base b representation of the positive integer n. For example, $S_3(24) = 4$ since $24 = 220_3$ and $2 + 2 + 0 = 4$.

(a) Prove that $S_b(n) = S_b(bn)$.
(b) Determine $S_b(1 + (b - 1)b^3)$.
(c) If $0 \le x < b^m$, evaluate $S_b(b^m - 1 - x) + S_b(x)$.
(d) Prove that n is divisible by $b - 1$ if and only if $S_b(n)$ is divisible by $b - 1$.

Chapter 3
Polynomials and Complex Numbers

In the first two chapters, we have focused our attention on numbers and numerals, with particular emphasis on the structure of various sets of numbers. In this chapter, we will begin to study functions, a concept that plays an important role in many branches of mathematics. For now, we'll restrict ourselves to a special kind of function—the polynomial.

We'll consider polynomials from two points of view. First, we'll treat them as a set of objects for which we can define arithmetic operations. We'll see (perhaps surprisingly) that the set of polynomials has much the same structure as the integers and, consequently, we can ask many of the same questions about polynomials—involving the Division Algorithm, unique factorization, and so on—that we asked about integers.

Next, we'll consider the solution of polynomial equations, that is, determining which numbers result in a value of 0 when substituted in the polynomial. In the process, we'll see that the set of real numbers (rationals plus irrationals) is not adequate; there are polynomial equations whose solutions are not real. As a result, we'll introduce the complex numbers.

Finally, we'll relate the study of polynomials to three famous problems in geometric constructions.

3.1. Polynomials

In your high school algebra courses, you undoubtedly encountered the term *polynomial* or *polynomial function*. For the purposes of this course, we'll restrict ourselves to polynomials in one variable. Some examples of polynomial functions are

$$f(x) = 1 + 2x - x^4 \quad \text{and} \quad g(x) = x^{31} + 2x^{28} - 5x^{17} + 12.$$

The largest exponent in the polynomial is called the *degree* of the polynomial; hence, f is of degree 4, and g is of degree 31. We'll assume that the coefficients of the polynomial are real numbers.

In the past, you have probably thought of polynomials as functions to be evaluated, analyzed, graphed, and solved. We'll do some of that in Section 3.3. However, polynomials (and any other type of function) can also be viewed as objects on which we can perform arithmetic operations. We can then determine under which operations the polynomials are closed. Depending on the answers to the closure questions, we can then investigate other properties, much as we did with integers and rationals.

   1. Complete the following definition.

Definition p is a polynomial function of degree n iff p can be expressed in the form $p(x) = $ _____ , for all x in its domain.

Note that the definition says that p is a polynomial if it *can be expressed* in a certain form. It is possible that p may be written in some other form that is equivalent to the desired form, at least for values of x in the domain. For example, the function $p(x) = (x + 3)(x^2 - 2x + 7)$ is not in the desired form, but it can be rewritten in the desired form by multiplying the two factors. This is similar to what we encountered in the definition of rational number—x is rational iff it can be expressed as $\frac{p}{q}$, where p and q are integers. Furthermore, the polynomial must be represented by the *same* formula over its entire domain. So, for example,

$$f(x) = \begin{cases} x^2 + 1, & x \geq 0 \\ 1 - x^3, & x < 0 \end{cases}$$

is not, by our definition, a polynomial.

Also note that the polynomial is p, not $p(x)$; $p(x)$ is the notation used to describe how the polynomial p acts on the variable x. Equivalently, $p(x)$ represents the numerical value obtained when x is substituted into p. The choice of the letter x is arbitrary. In other words, $p(x) = x^2$, $p(y) = y^2$, and $p(\xi) = \xi^2$ all represent the same polynomial p.

   2. Which of the following are polynomial functions? For each one that is not a polynomial function, explain why.

   (i) $p(x) = x^2 + 4x + 19$

(ii) $p(x) = x^2 + 4\sqrt{x} + 19$

(iii) $p(x) = \dfrac{1}{x^3}, x \neq 0$

(iv) $p(x) = \dfrac{x^3 - 1}{x - 1}, x \neq 1$

(v) $p(x) = \cos(2 \arcos(x)), -1 \leq x \leq 1$

(vi) $p(x) = e^x$

[*Hint: For (v), note that* $\cos(2\theta) = 2 \cos^2(\theta) - 1$.]

Now let's define the usual arithmetic operations on polynomials.

Definition   Let p and q be polynomials.

The *sum* of p and q is the polynomial $p + q$ defined by $(p + q)(x) = p(x) + q(x)$.

The *product pq* is defined by $(pq)(x) = p(x)q(x)$.

The *quotient $\frac{p}{q}$* is defined by $(\frac{p}{q})(x) = \frac{p(x)}{q(x)}$, for all x such that $q(x) \neq 0$.

Notice that since $p(x)$ and $q(x)$ are real numbers, then we have defined the sum (and product and quotient) of two polynomials in terms of the corresponding operations on real numbers. Although technically we have not defined such operations on real numbers (in fact, we haven't defined the real numbers either), we shall assume they can be appropriately defined and that the real numbers are closed under all operations.

3. Determine whether the set of polynomials is closed under each of the following operations. If you answer yes, prove it.

(i) *Addition.*
(ii) *Multiplication.*
(iii) *Division.*

4. If p and q are polynomials of degree m and n, respectively, what is the degree of $p + q$? What is the degree of pq?

The real number 0 (which, conveniently, is also rational and an integer) has the property that $x + 0 = x$, for all x. The corresponding polynomial is the *zero polynomial* $z(x) = 0$ for all x. Its graph is the x-axis. A problem arises when we try to define the degree of the zero polynomial. The definition of polynomial given earlier doesn't really cover this case; it defines the degree as the "largest exponent" on a term with nonzero coefficient, but the zero polynomial has no terms with nonzero coefficient.

This is an example of a situation that occurs frequently in mathematics. We have a definition of something that applies in all but one special case. We are free to define the special case in such a way that other known results apply to that special case. We often use the phrase "by convention" to describe this seemingly arbitrary choice. For instance, the well-known factorial function is defined by $n! = n(n - 1)(n - 2) + \cdots + (2)(1)$, if n is a positive integer. We also define $0! = 1$ to make certain results true for $n = 0$. For example, the coefficient of x^n in the Taylor series expansion of a function f is

$$\frac{f^{(n)}(0)}{n!},$$

where $f^{(n)}(0)$ is the nth derivative of f, evaluated at $x = 0$. To make this true for $n = 0$, we need to have $0! = 1$. We'll see other examples in Chapter 4.

Here, we'd like to define the degree of the zero polynomial in a way that maintains the truth of your answers to question 4. One convention is to set the degree of the zero polynomial equal to $-\infty$. We also have to claim that $-\infty + x = -\infty$, for every (finite) positive number x.

5. *Show that this convention makes your answers to question 4 true.*

6. *What is the degree of constant polynomials other than the zero polynomial— that is, $p(x) = k$, where $k \neq 0$?*

3.1.1. Division and Factoring

We have shown that the set of polynomials is closed under addition and multiplication but not under division. This means that the polynomials have a structure similar to the integers. Thus, it is reasonable to ask if the polynomials have properties such as the Division Algorithm and unique factorization.

Recall that the Division Algorithm for integers says that whenever we divide two integers a and b, we get a unique quotient q and remainder r, where $0 \leq r < b$. It is the restriction on r that ensures the uniqueness.

Now suppose f and g are polynomials. There are two questions that must be answered:

- Do we always get a polynomial quotient and a polynomial remainder when we divide f by g?
- What condition on the remainder ensures that the quotient and remainder are unique?

7. *Make up several division problems involving polynomials of various degrees. In each case, note the quotient and remainder. Are the quotients and remainders polynomials? How does the degree of the remainder relate to the degree of the divisor?*

8. Complete the following theorem.

Theorem 3.1 (Division Algorithm for Polynomials)   Given two polynomials f and g, there exist unique polynomials q and r such that $f(x)$ = _____, where _____.

In the event that $g(x) = x - b$ is a linear divisor, then the remainder must be constant. Moreover, we have the following corollary.

Corollary 3.1   The remainder when $f(x)$ is divided by $g(x) = x - b$ is $f(b)$.

9. Prove Corollary 3.1.

Now let's think about factoring. When we factor an integer n, we express it as the product of two *integers*, each of which is *strictly smaller than n*. Thus, 24 can be factored as 8×3 or 12×2 or 4×6, but not as 24×1 or $48 \times \frac{1}{2}$. Integers that cannot be factored are said to be prime.

To factor a polynomial p, we must write it as the product of two other polynomials, f and g.

10. What condition must be imposed on f and g to make the factoring "proper" (analogous to requiring the factors of an integer to be strictly smaller than the integer)?

11. What would happen if we didn't require the factors themselves to be polynomials?

12. Argue that neither factor can be of degree 0.

If a polynomial cannot be factored, it is said to be *irreducible*.

13. Prove that every linear (first-degree) polynomial is irreducible.

There is one more issue that must be settled before we can go on. Strictly speaking, when talking about polynomials, we must specify the set of numbers from which the coefficients are drawn. At the beginning of this chapter, we assumed that the coefficients were real numbers. In some cases, we may wish to change this assumption; indeed, we could restrict the coefficients to rational numbers, integers, or just elements of Z_m (see Section 1.1), or we could expand the set of coefficients to include the complex numbers.

The reason this is important is that there are polynomials that are irreducible if the coefficients are, say, rational, but factorable if the coefficients are real. In this case, we'll say that the polynomial is "factorable over **R,** but not factorable over **Q.**" For example, $x^2 - 2 = (x - \sqrt{2})(x + \sqrt{2})$ is the product of two polynomials with real, but not rational, coefficients. Similarly, $x^2 + 1 = (x - i)(x + i)$ is factorable over **C** (the complex numbers), but not over **R**.

Factoring polynomials is not always an easy task, particularly if the polynomial is of degree 4 or more. We are aided in the search for linear factors by the following theorem.

Theorem 3.2 (Factor Theorem)   If p is a polynomial and b is a number such that $p(b) = 0$, then $p(x) = (x - b)q(x)$, where q is a polynomial.

We say that b is a *zero* of the polynomial p, or that b is a *root* of the polynomial equation $p(x) = 0$.

14. *Factor each of the following over* **Q**, *over* **R**, *and over* **C**.

   (i) $x^2 + 6x - 7$ (ii) $x^3 + 6x - 7$ (iii) $x^4 - 6x^2 - 7$

15. *If p is factorable over* **Q**, *is it factorable over* **R**? *Is the converse true?*

16. *Let* $p(x) = ax^2 + bx + c$ *be a quadratic polynomial. For what values of a, b, and c is p factorable over* **Q**? *over* **R**?

17. *Show that every cubic polynomial is factorable over* **R**. *(Think about its graph.)*

The next question is whether the Fundamental Theorem of Arithmetic can be adapted to polynomials. In other words, can every polynomial be expressed *uniquely* as a product of polynomials? We'll restrict our discussion to factoring over the integers **Z**.

At first glance, it would appear that the answer is no since, for example, $4x^2 - 16 = (2x - 4)(2x + 4) = (x - 2)(4x + 8)$. All factors are irreducible according to our definition. However, a closer look is in order.

Suppose we factor out the constant 4 first: $4x^2 - 16 = 4(x - 2)(x + 2)$. Each of the factors $x - 2$ and $x + 2$ is irreducible. Moreover, they have no constant factors other than $+1$ and -1. The only other factorizations of $4x^2 - 16$ must contain $x - 2$ and $x + 2$ or $-(x - 2)$ and $-(x + 2)$.

Let's agree that, when factoring polynomials with coefficients in **Z**, we first factor out the largest possible constant factor. This factor is called the *content* of the polynomial.

18. *How is the content of a polynomial related to its coefficients?*

After factoring out the content of the polynomial, the remaining polynomial has no constant integer factor. Such polynomials are said to be *primitive*. So, for example, $x - 2$ is primitive, but $2x - 4$ is not primitive.

19. *Prove that if a primitive polynomial is reducible, then its factors must be primitive.*

We are now ready to state Theorem 3.3.

Theorem 3.3   Let p be a polynomial of degree greater than 0 with coefficients in **Z**. Then there exist irreducible, primitive polynomials $p_1(x), \ldots, p_r(x)$ such that $p(x) = kp_1(x)p_2(x) \cdots p_r(x)$, where k is the content of the polynomial. Moreover, this factorization is unique except for the order of the factors and for factors of $+1$ or -1.

The constant k is itself a polynomial of degree 0. It is not considered irreducible or reducible, in the same way that the integer 1 is not considered prime or composite. When studying sets of objects that have a structure similar to the integers, we must identify those elements that are neither "prime" nor "composite." For the positive integers, the only such element is 1; for polynomials with integer coefficients, those elements are all the constant polynomials. Once we identify those elements, then we can talk about unique factorization.

3.1.2.   The Eisenstein Irreducibility Criterion

As we said earlier, determining whether a polynomial is reducible is not an easy task, except for quadratics and cubics. There are, however, some theorems that can help us. One such theorem is the Eisenstein Irreducibility Criterion (EIC, for short).

Theorem 3.4 (EIC)   Let $f(x) = a_n x^n + a_{n-1} x^{n-1} + \cdots + a_0$ be a polynomial with integer coefficients. If there exists a prime integer p such that:

    (i) p is not a divisor of a_n,
    (ii) p is a divisor of every other coefficient $a_{n-1}, \ldots, a_1, a_0$, and
    (iii) p^2 is not a divisor of a_0,

then f is irreducible over **Z**.

   20. *Use the EIC to show that $2x^4 + 21x^3 - 6x^2 + 9x - 3$ is irreducible.*

   21. *Show that $x^2 + x + 1$ does not satisfy the EIC for any prime, but it is still irreducible.*

   22. *Determine a cubic polynomial that EIC implies is irreducible over* **Z.**

Although it is not hard to prove the EIC in general, we shall just consider the case of cubic polynomials. The proof is by contradiction.
   Let $f(x) = a_3 x^3 + a_2 x^2 + a_1 x + a_0$. Assume f is reducible over **Z**. We'll show that there cannot be a prime p that satisfies (i), (ii), and (iii) in Theorem 3.4.

   23. *If f is reducible, then $f(x) = (b_1 x + b_0)(c_2 x^2 + c_1 x + c_0)$. Why is this the only possibility?*

24. Write equations expressing each coefficient of f as a combination of the b's and c's.

25. Argue that p must divide either b_0 or c_0, but not both.

26. Assume that p divides c_0. Argue that it must also divide c_1 and c_2.

27. Why is this a contradiction? (If we had assumed p divides b_0, then it would also have to divide b_1. In either case, we get the same contradiction.)

3.2.   Complex Numbers

As we have seen, polynomials with real coefficients may have factors that involve complex (nonreal) numbers. Put differently, a real polynomial may have nonreal zeroes. Thus, a discussion of complex numbers and their properties is in order.

A *complex number* is of the form $z = x + yi$, where x and y are real and $i = \sqrt{-1}$. If $x = 0$, then z is purely imaginary. If $y = 0$, then z is real. Complex numbers can be plotted on a Cartesian plane with the horizontal axis corresponding to the real part and the vertical axis corresponding to the imaginary part. Thus, the number $z = x + yi$ corresponds to the point whose coordinates are (x, y). Complex numbers can also be expressed in *polar form*: $z = r[\cos(\theta) + i \sin(\theta)]$, where r is called the *magnitude* of z, denoted $\|z\|$, and θ is called the *argument* of z, denoted $\arg(z)$. We often abbreviate this as $r \operatorname{cis}(\theta)$. [cis is a symbol for $\cos + i \sin$]. Note that θ is measured counterclockwise from the positive x-axis and we agree that θ is between 0 and 2π.

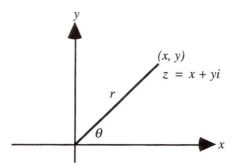

We'll agree that two complex numbers $z = x + yi$ and $w = u + vi$ are equal iff $x = u$ and $y = v$.

1. Express r and θ in terms of x and y.

2. Where on the Cartesian plane are all the complex numbers whose magnitude is 1? Where are all the numbers whose argument is $\frac{\pi}{4}$?

We can perform arithmetic operations with complex numbers, but first we have to define those operations, much as we did with polynomials. Again, those operations will be defined in terms of the corresponding operations on real numbers.

3. Let $z = x + yi$ and $w = u + vi$. Define the sum $z + w$. Illustrate with a few examples of your choice.

The sum of two complex numbers can be represented geometrically, as shown below. The line joining the origin with $z + w$ is actually the diagonal of a parallelogram whose sides are the lines joining the origin to z and w, respectively.

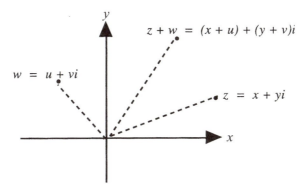

4. Define the product zw. Illustrate with a few examples of your choice.

5. If $z = r_1 \operatorname{cis}(\theta_1)$ and $w = r_2 \operatorname{cis}(\theta_2)$, show that $zw = r_1 r_2 \operatorname{cis}(\theta_1 + \theta_2)$. [You will have to recall or look up some formulas about the cosine and sine of the sum of two angles.]

The product of two complex numbers can be represented geometrically, as shown below.

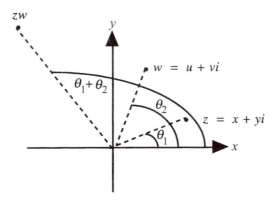

The length of the line segment joining the origin to zw is the product of the lengths of the other two segments. The angle between the line segment to zw and the x-axis is the sum of the other two angles.

6. Use the result of question 5 to complete and prove the following theorem, known as DeMoivre's Theorem.

Theorem 3.5 (DeMoivre's Theorem)   Let $z = r \operatorname{cis}(\theta)$ be a complex number and let n be a positive integer. Then $z^n = $ _____.

7. Suppose $z = 1 + i$. Use DeMoivre's Theorem to compute z^6.

8. For which values of n will $(1 + i)^n$ be a real number?

9. Expand $(\cos(\theta) + i \sin(\theta))^3$. Then use DeMoivre's Theorem to express $\cos(3\theta)$ and $\sin(3\theta)$ in terms of $\cos(\theta)$ and $\sin(\theta)$.

Now let's see how the magnitudes of $z + w$ and zw are related to the magnitudes of z and w.

10. Try a few numerical examples to see if you can determine a relationship between $\|z + w\|$ and $\|z\| + \|w\|$. Prove your conjecture.

11. Conjecture and prove a relationship between $\|zw\|$ and $\|z\| \|w\|$.

12. Prove that $\|z^n\| = \|z\|^n$, for all n.

Before we can discuss division, we need the following definition.

Definition   The *complex conjugate* of z is $\bar{z} = x - yi$.

13. How are the magnitudes of z and $\bar{z}$ related? How are their arguments related? Show that $\bar{z} = r \operatorname{cis}(-\theta)$. Given a graph depicting a complex number z, how would you locate $\bar{z}$?

14. What kind of number is $z\bar{z}$?

15. Now consider the quotient $q = \dfrac{z}{w} = \dfrac{x + yi}{u + vi}$. Show that q is a complex number by expressing it in the form $a + bi$, where a and b are real.

3.2.1.   *n*th Roots of One

Consider the equation $x^n = 1$, where n is a positive integer.

16. Argue that if n is odd, then there is exactly one real root. What is it?

17. Argue that if n is even, then there are exactly two real roots. What are they?

18. By factoring and using the quadratic formula, completely solve the case n = 3.

19. Express the nonreal roots in question 18 in polar form. Verify that they are correct by using DeMoivre's Theorem. Plot the roots on the Cartesian plane.

20. Show that one nonreal root is the reciprocal of the other.

21. Now completely solve the case n = 4.

22. Express the roots of $x^5 = 1$ in polar form and plot on the Cartesian plane.

It is not obvious how to express the nonreal roots in question 22 in "radical" form, as we could in question 18. If we try factoring the original equation, we get $x^5 - 1 = (x - 1)(x^4 + x^3 + x^2 + x + 1)$. While it is possible to solve the resulting fourth-degree equation, we shall not do so here. We'll look at a different approach in the Additional Questions at the end of this chapter.

3.3. Roots of Polynomials

For every polynomial function p, there is a corresponding polynomial equation $p(x) = 0$. We will use the phrase "r is a *root* of a polynomial equation" or "r is a *zero* of a polynomial" interchangeably.

1. What does it mean to say that "r is a root of a polynomial equation"?

2. What is the connection between the zeroes of a polynomial and its graph?

3. If p is a polynomial of degree n, how many roots does p(x) = 0 have?

4. Suppose the zeroes of an nth-degree polynomial p(x) are $r_1, r_2, \ldots, r_n$. Express p(x) as a product of linear factors.

5. Suppose p(x) = u(x)v(x). Show that every zero of p must either be a zero of u or a zero of v.

In general, finding the roots of a polynomial equation is not an easy task. If the polynomial is factorable as the product of other polynomials, then we may be able to use the results of question 5 to find the roots. However, most polynomials are not factorable, and we are forced to use other methods.

We'll begin with the simplest case—quadratic (second-degree) polynomials. Let $p(x) = ax^2 + bx + c$ be a quadratic polynomial. The well-known quadratic formula tells us that the zeroes of this polynomial are given by

$$x = \frac{-b \pm \sqrt{b^2 - 4ac}}{2a}.$$

To prove this, we need to rewrite the polynomial in a different, but algebraically equivalent, form.

6. *Show that* $p(x) = a\left(x + \dfrac{b}{2a}\right)^2 + c - \dfrac{b^2}{4a}.$

7. *Use this form to solve* $p(x) = 0$. *Do you get the quadratic formula?*

Depending on the values of a, b, and c, the roots may be real or imaginary.

8. *Under what conditions are the roots real? imaginary? equal? Draw a graph of a typical polynomial in each case. Create an example of each case.*

9. *Express the sum and product of the roots in terms of* a, b, *and* c.

10. *Show that the sum of the reciprocals of the roots is* $\dfrac{-b}{c}$.

11. *Show that, if neither* a *nor* c *is* 0, *the roots of* $cx^2 + bx + a = 0$ *are the reciprocals of the roots of* $ax^2 + bx + c = 0$.

12. *Prove that if* a, b, *and* c *are odd, then the roots cannot be rational.*

Now we'll look at cubic (third-degree) polynomials. Let $p(x) = ax^3 + bx^2 + cx + d$ be a cubic polynomial. We expect that p has three zeroes, which may be real or imaginary. Let r_1, r_2, and r_3, be the zeroes.

13. *Show that the sum of the roots of* $p(x) = 0$ *is* $\dfrac{-b}{a}$ *and the product of the roots is* $\dfrac{-d}{a}$. *Determine an algebraic combination of the roots that is equal to* $\dfrac{c}{a}$ *(or, perhaps,* $\dfrac{-c}{a}$*).*

14. *Argue that* p *must always have at least one real zero.*

15. *Show that since* a, b, c, *and* d *are real, then any imaginary roots must occur as conjugate pairs.*

16. *Construct a cubic polynomial that has exactly one real zero.*

17. *Construct a cubic polynomial that has three real zeroes.*

Note that when we say that p has three real zeroes, we do not necessarily mean that they are distinct. If p has three real zeroes, it is possible that two or more of them are equal.

18. *(Note: Questions 18 and 19 require calculus.) Assume $a > 0$. Certainly p will have only one real zero if it is a strictly increasing function. Derive a relationship between the coefficients of p that ensures that p is strictly increasing. Is it necessary that p be strictly increasing in order for it to have exactly one real zero?*

19. *Let $p(x) = x^3 - 3x^2 - 9x + d$. Determine all values of d such that p has exactly one real zero.*

There are formulas for solving cubic equations and quartic (fourth-degree) equations, but they are quite complicated. Most interestingly, it has been proven that there cannot be a formula involving ordinary algebraic operations that can be used to solve quintic (fifth-degree) or higher-degree polynomials.

3.3.1. Descartes' Rule of Signs

As we said earlier, there are no methods that can be used to solve polynomial equations in general. However, we can sometimes get some information about the roots by inspecting the coefficients of the polynomial.

20. *What are the roots of $x^2 + 1 = 0$? What are the roots of $x^4 + 5x^2 + 4 = 0$?*

21. *Does $x^6 + 8x^4 + 7x^2 + 13 = 0$ have any real roots? Explain.*

22. *The polynomials in questions 20 and 21 have something in common. What is it and how does it affect the nature of the roots? Prove your answer.*

23. *Show that a polynomial with all positive coefficients can't have any positive zeroes.*

24. *What conditions on the coefficients would ensure that the polynomial would have no negative zeroes?*

The reason that a polynomial p with all positive coefficients can't have any positive zeroes is that $p(x)$ will consist of all positive terms when $x > 0$. There are no negative terms to "cancel out" the positive terms. Thus, it seems that the existence of positive roots of a polynomial depends on whether there are terms with both positive and negative coefficients. This leads us to the following definition.

Definition The number of *sign changes* in a polynomial is the number of times the sequence of coefficients (read left to right) changes either from positive to negative or negative to positive.

So, for example, $x^3 - 3x^2 - 12x + 7$ has two sign changes and $x^4 + 9x^3 - 5x^2 + 12x - 17$ has three sign changes.

25. Show that a polynomial with no sign changes has no positive zeroes.

26. Let $p(x) = ax + b$. Show that the zero of p is positive if and only if there is one sign change.

Let $p(x) = ax^2 + bx + c$. The pictures below illustrate five different possibilities for the graph of p.

27. For each graph, determine the number of sign changes in p and the number of positive zeroes of p. How does the number of positive zeroes of p correspond to the number of sign changes?

28. For negative zeroes, consider the polynomial $q(x) = p(-x)$, where $p(x) = ax^2 + bx + c$. Clearly, every positive zero of q corresponds to a negative zero of p. Thus, relate the number of negative zeroes of p to the number of sign changes in q.

29. Create a cubic polynomial p whose zeroes are specified in each case below. For each one, count the number of sign changes of $p(x)$ and $p(-x)$ and relate it to the number of positive and negative zeroes.

(i) three positive zeroes	*(ii) two positive, one negative*
(iii) one positive, two negative	*(iv) three negative*
(v) one positive, two imaginary	*(vi) one negative, two imaginary*

30. Complete the following theorem, which is known as Descartes' Rule of Signs.

Theorem 3.6 (Descartes' Rule of Signs) The number of positive zeroes of a polynomial p is _____ the number of sign changes in $p(x)$. The number of negative zeroes is _____.

The proof is by induction on the degree of p. However, it is somewhat more complicated than other induction proofs, so we won't discuss it here.

31. Use Descartes' Rule of Signs to characterize the zeroes of $x^{11} + x^8 - 3x^5 + x^4 + x^3 - 2x^2 + x - 2$. What is the minimum number of nonreal zeroes?

32. Create a few more examples of polynomials of degree 4 or higher. Use Descartes' Rule of Signs to obtain information about the roots. Verify by graphing the polynomial on a calculator or computer.

3.3.2. Rational Roots

Let p be a polynomial with integer coefficients. We'll prove in this section that there are only a finite number of possibilities for *rational* zeroes of p.

Theorem 3.7 Let $p(x) = a_n x^n + a_{n-1} x^{n-1} + \cdots + a_1 x + a_0$, where all a_j are integers. If $x = \frac{r}{s}$ is a rational zero of p, where r and s are relatively prime integers, then r is a divisor of a_0 and s is a divisor of a_n.

If $x = \dfrac{r}{s}$ is a zero of p, then

$$p\left(\frac{r}{s}\right) = a_n\left(\frac{r}{s}\right)^n + a_{n-1}\left(\frac{r}{s}\right)^{n-1} + \cdots + a_1\left(\frac{r}{s}\right) + a_0 = 0$$

or, upon multiplying by s^n,

$$a_n r^n + a_{n-1} r^{n-1} s + \cdots + a_1 r s^{n-1} + a_0 s^n = 0.$$

33. Rewrite this as $-(a_n r^n + a_{n-1} r^{n-1} s + \cdots + a_1 r s^{n-1}) = a_0 s^n$ and conclude that a_0 must be divisible by r.

34. By a similar argument, show that a_n is divisible by s.

35. List all possibilities for rational roots of

$$6x^4 - 7x^3 + 8x^2 - 7x + 2 = 0.$$

Which ones actually are roots? How many nonrational roots are there?

Theorem 3.7 provides a convenient way of proving some numbers are irrational. For example, consider the polynomial equation $p(x) = x^2 - 2 = 0$. According to the theorem, the only possible rational roots are ± 1 and ± 2. Simple substitution eliminates all of these; hence, all real roots of the equation must be irrational. Since $\sqrt{2}$ is one of the real roots, then $\sqrt{2}$ must be irrational.

36. Let n and k be positive integers. Prove that if $\sqrt[n]{k}$ is not an integer, then it must be irrational.

We can use the same approach to show that more "complicated" numbers such as $\sqrt{2} + \sqrt{3}$ are irrational. (Remember, the irrationals are not closed under addition, so this is not a trivial result.)

37. Show that $x = \sqrt{2} + \sqrt{3}$ is a root of the polynomial equation $x^4 - 10x^2 + 1 = 0$.

38. What are the only possible rational roots of this equation?

39. Conclude that $\sqrt{2} + \sqrt{3}$ must be irrational.

40. Use a similar argument to show that $\sqrt{2 + \sqrt{5}}$ is irrational.

Irrational numbers such as $\sqrt{2} + \sqrt{3}$ and $\sqrt{2 + \sqrt{5}}$ are roots of polynomials with integer coefficients. There are, however, irrational numbers, such as π, which are not roots of polynomials with integer coefficients. This leads to the following definition.

Definition   A real number x is said to be *algebraic* if there exists a polynomial p with integer coefficients such that $p(x) = 0$. Real numbers that are not algebraic are said to be *transcendental*.

Proving that there are transcendental numbers is not too difficult. The proof involves showing that the set of algebraic numbers is, in some sense, "smaller" than the set of real numbers; consequently, there must be some real numbers that are not algebraic. On the other hand, proving that a particular number such as π is transcendental is much more difficult.[1]

41. Prove that the rational numbers are algebraic.

42. Prove that every number of the form

$$\frac{a + b\sqrt{c}}{d}, \quad \text{where a, b, c, and d are integers with } c > 0,$$

is algebraic.

In fact, any real number that can be computed by a finite number of algebraic operations—addition, multiplication, division, and root-taking—with integers is algebraic. There are, however, algebraic numbers that do not appear to be computed in this manner. Here's an example.

---

[1] Charles Hermite (1822–1901) proved that e is transcendental in 1873. Ferdinand Lindemann (1852–1939) proved that π is transcendental in 1882. Both proofs made use of techniques developed by Joseph Liouville (1809–1882), who is also known for his work in complex analysis.

According to DeMoivre's Theorem,

$$\cos(3\theta) + i \sin(3\theta) = [\cos(\theta) + i \sin(\theta)]^3$$
$$= \cos^3(\theta) + 3i \cos^2(\theta)\sin(\theta) - 3 \cos(\theta)\sin^2(\theta) - i \sin^3(\theta).$$

By equating the real parts of this equation, we get

$$\cos(3\theta) = \cos^3(\theta) - 3 \cos(\theta) \sin^2(\theta)$$
$$= \cos^3(\theta) - 3 \cos(\theta)(1 - \cos^2(\theta))$$
$$= 4 \cos^3(\theta) - 3 \cos(\theta).$$

Now let $\theta = 20°$ and $x = \cos(\theta)$.

43. *Show that* $8x^3 - 6x - 1 = 0$ *and thus conclude that* $\cos(20°)$ *is algebraic.*

44. *Show that* $\sin(10°)$ *is algebraic.*

In a somewhat similar manner, we can show that any trigonometric function of any *rational* multiple of 180° is algebraic. In some cases, it is relatively simple to express the values of these trigonometric functions in terms of algebraic operations. For example, $\sin(60°) = \dfrac{\sqrt{3}}{2}$ or $\tan(15°) = \dfrac{\sqrt{3} - 1}{\sqrt{3} + 1}$ (obtained from a half-angle formula for the tangent). See question 15 at the end of the chapter for another example.

Trigonometric functions of other angles are transcendental, as are the many values of logarithmic and exponential functions. Indeed, these functions are often referred to as *transcendental functions*.

3.4. Geometric Constructions

In your high school geometry course, you probably spent some time doing a variety of constructions using an unmarked straightedge and compass. Some of the problems that can be solved in this manner include constructing the perpendicular bisector of a line segment, the bisector of an angle, a line parallel to a given line through a given point, a circle through three given points, and a triangle given the lengths of its sides.

All of these boil down to combinations of finding the intersections between two lines, two circles, or one circle and one line. For example, to construct the perpendicular bisector of a line segment, we draw circles with each end of the segment as centers and radius greater than half the length of the segment. Then we use the straightedge to connect the points of intersection of the circles.

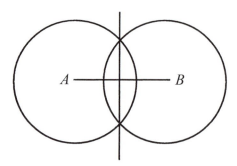

In this section, we're going to look at a slightly different question: Suppose we are given a line segment of length 1. For what values of r can we construct a line segment of length r?

We shall see that there is a set of numbers that can be constructed using the allowable tools. We call these numbers *constructible* (obviously). It turns out that the set of constructible numbers includes the rationals and some, but not all, irrationals. In fact, it doesn't even include all the algebraic numbers, although every constructible number is algebraic.

Once we decide which numbers are constructible, we'll be able to show that three very famous problems that had plagued mathematicians since the time of the ancient Greeks are impossible.

PROBLEM 1: Duplicating the cube

Given a cube of volume V, construct a cube of volume $2V$.

PROBLEM 2: Trisecting an angle

Given an arbitrary angle of size θ, construct an angle of size $\frac{\theta}{3}$. (Note the word *arbitrary*. Some specific angles, such as π, are trisectible. We want a method for trisecting every angle.)

PROBLEM 3: Squaring the circle

Given a circle of area A, construct a square whose area is also A.

Throughout this discussion, it is important to remember that we are given a line segment whose length *we define* to be 1 unit. All other line segments are measured in terms of that unit.

1. Describe how to construct a line segment of length n, where n is an integer.

2. Given line segments of length x and y, describe how to construct line segments of lengths x + y and x − y.

We can also construct segments of length xy and $\frac{x}{y}$, but this takes a little more work. We use the fact that if AB is parallel to CD, then $PA{:}PC = PB{:}PD$. So we begin by drawing two lines that intersect at *P*. On one line, we mark off point *A* so that *PA* = 1. Then we mark off point *C* so that *PC* = *x*. On the other line, we mark off point *B* so that *PB* = *y*. We draw *AB*, and then draw *CD* parallel to *AB*.

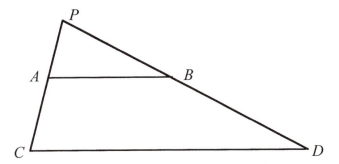

3. Which segment has length xy?

4. Modify the construction to produce a segment of length $\frac{x}{y}$.

5. Argue that all rational numbers are constructible.

In addition to the rational operations, we can also construct $\sqrt{x}$, where *x* is itself constructible. The construction is depicted in the diagram below. First we draw a line and locate points *A*, *P*, and *B* such that *AP* = *x* and *PB* = 1. Next we draw a circle with *AB* as diameter, and then construct a perpendicular to *AB* at *P*. The perpendicular intersects the circle at *C*.

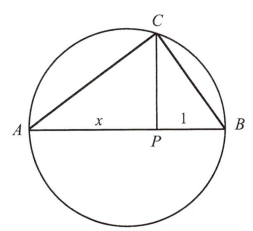

6. Argue that triangles ACP and BCP are similar. Use this to prove that CP = $\sqrt{x}$.

So, we can construct the sum, difference, product, and quotient of two numbers and the square root of a number. Moreover, these are the *only* operations we can construct, a fact that takes some proof. Of course, we can repeat the operations; hence, a fairly complicated looking number such as

$$\sqrt{\frac{1+\sqrt{11}}{3+\sqrt{7}}}$$

is constructible. It follows that all constructible numbers are algebraic.

The converse of this statement is false; that is, there are algebraic numbers that are not constructible. To characterize which ones are, we need the following definition.

Definition   An algebraic number x is of *degree n* if there exists a polynomial p of degree n with integer coefficients such that $p(x) = 0$ and n is the smallest integer for which this is true.

For example, $\sqrt{2}$ is of degree 2 since $\sqrt{2}$ is a zero of $p(x) = x^2 - 2$ but $\sqrt{2}$ is not a zero of any first-degree polynomial. Now consider $\sqrt[3]{2}$. We know that $\sqrt[3]{2}$ is a root of the third-degree polynomial $p(x) = x^3 - 2 = 0$; hence, $\sqrt[3]{2}$ is of degree ≤ 3. We'd like to show that it is of degree (exactly) 3, by showing that there is no polynomial of lower degree of which $\sqrt[3]{2}$ is a zero. We'll assume that we've already shown that $\sqrt[3]{2}$ is irrational.

7. Show that $\sqrt[3]{2}$ is not a root of any first-degree polynomial; that is, show that there are no integers a and b such that $a\sqrt[3]{2} + b = 0$.

We also have to show that $\sqrt[3]{2}$ is not the root of any quadratic polynomial. In other words, there are no integers a, b, and c such that $a(\sqrt[3]{2})^2 + b\sqrt[3]{2} + c = 0$ or, equivalently, $a\sqrt[3]{4} + b\sqrt[3]{2} = -c$.

8. Square both sides of the equation above and rearrange terms to obtain $b^2\sqrt[3]{4} + 2a^2\sqrt[3]{2} = c^2 - 4ab$.

9. Use the last two equations to show that

$$\sqrt[3]{2} = \frac{4a^2b - ac^2 - b^2c}{b^3 - 2a^3}$$

provided $b^3 - 2a^3 \neq 0$. Why is this a contradiction?

10. Why is there also a contradiction if $b^3 - 2a^3 = 0$?

Hence, we've shown that $\sqrt[3]{2}$ is an algebraic number of degree 3.

Determining the degree of an algebraic number is not always a simple process, and the "logical" answer may not be right. For example, we may be led to believe that the number $\sqrt{2} + \sqrt{3}$ is of degree 2 since it has square roots in it. In fact, this number is of degree 4 since it is a root of $x^4 - 10x^2 + 1 = 0$ (see question 37 of Section 3.3) and not of any lower-degree polynomial (a fact that takes some work to prove).

Now for the main result, which we state without proof.

Theorem 3.8   A real number x is constructible iff it is algebraic of degree n, where $n = 2^k$ for some positive integer k.

Thus, $\sqrt{2}$, $\sqrt{2} + \sqrt{3}$, and $\sqrt{1 + 3\sqrt{7}}$ are constructible, but $\sqrt[3]{2}$ is not constructible. It also follows that no transcendental number is constructible.

We can now easily dispense with the three famous problems stated earlier.

11. Show that doubling the cube requires constructing a line segment of length $\sqrt[3]{2}$.

Since this is not constructible, then it is impossible to double the cube. For the angle trisection problem, it suffices to show that 60° is not trisectible. In other words, we have to show that it is impossible to construct an angle of 20°. In right triangle ABC, assume that angle $BAC = 60°$, angle $DAC = 20°$, and $AC = 1$. It follows that $AD = \sec(20°)$. So, to construct the angle of 20°, we'd have to construct a line segment of length $\sec(20°)$. Since the constructible numbers are closed under division, we could do this if we could construct a segment of $\cos(20°)$.

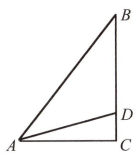

In the last section, we showed that $\cos(20°)$ is algebraic and satisfies the cubic equation $8x^3 - 6x - 1 = 0$. It takes some work to show that there is no lower-degree polynomial of which $\cos(20°)$ is a zero. Hence, $\cos(20°)$ is a third-degree algebraic number which, by Theorem 3.8, is not constructible. Thus, it is impossible to find a method of trisecting an arbitrary angle.

Finally, we tackle the squaring the circle problem.

12. *Show that squaring the circle involves constructing a line segment of length $\sqrt{\pi}$. Why is it impossible to square the circle?*

3.5.   Additional Questions

1.   Let r and s be the roots of $ax^2 + bx + c = 0$. Derive expressions for each of the following in terms of a, b, and c.

(a) $r^2 + s^2$ (b) $r^3 + s^3$ (c) $\left(r + \dfrac{1}{s}\right)\left(s + \dfrac{1}{r}\right)$

2.   Determine the range of the rational function $f(x) = \dfrac{x^2 + x + 2}{3x + 1}$.

3.   Determine all polynomials p such that $p(x^2) = [p(x)]^2$.

4.   Let $p(x) = a_n x^n + a_{n-1} x^{n-1} + \cdots + a_0$ and $q(x) = a_0 x^n + a_1 x^{n-1} + \cdots + a_n$, where a_n and a_0 are nonzero. Prove that zeroes of p are the reciprocals of the zeroes of q.

5.   Determine all values of k such that $x^2 + kx + 1 = 0$ and $x^2 + x + k = 0$ have at least one root in common.

6.   The roots of $x^4 - ax^3 + ax^2 + bx + c = 0$ are r, s, t, and u. Find the minimum possible value of $r^2 + s^2 + t^2 + u^2$.

7.   Find the sum of the coefficients of the polynomial obtained after expanding $(1 - 3x + 3x^2)^{743}(1 + 3x - 3x^2)^{744}$.

8.   A *reciprocal polynomial* is one whose coefficients are symmetric about the middle. For example, $2x^2 + 5x + 2$ and $2x^3 - 5x^2 - 5x + 2$ are reciprocal polynomials.

(a) Show that $x = 0$ is never a zero of a reciprocal polynomial.

(b) Show that $x = -1$ is always a zero of a reciprocal polynomial of odd degree. Also, show that if p is a reciprocal polynomial of odd degree, then $p(x) = (x + 1)q(x)$, where q is also a reciprocal polynomial.

(c) Show that if $x = r$ is a zero of a reciprocal polynomial, then $x = \dfrac{1}{r}$ is also a zero.

(d) Solve the fourth-degree reciprocal polynomial equation $2x^4 + 5x^3 + x^2 + 5x + 2 = 0$. [*Hint:* First divide by x^2 and regroup terms. Then make the substitution $y = x + \dfrac{1}{x}$. You should get a quadratic in y.]

9.   Let p be a polynomial whose leading coefficient (i.e., the coefficient of the highest-power term) is 1. Show that if x is a rational zero of p, then x must be an integer. [*Hint*: Assume $x = \frac{r}{s}$, where r and s are integers with no common factor. Show that $s = 1$.]

10. Express all sixth roots of 1 in the form $a + bi$ for some a and b.

11. Express both square roots of i in the form $a + bi$ for some a and b.

12. Prove that the conjugate of the product of two complex numbers is equal to the product of their conjugates.

13. Show that the cubic equation $x^3 + px + q = 0$ has 3 real roots iff $D \le 0$, where $D = 27q^2 + 4p^3$. [*Hint*: Use calculus.]

14. Suppose the cubic equation $x^3 + px + q = 0$ has a nonreal root $a + bi$, where a, b, p, and q are all real and $q \ne 0$. Show that $aq > 0$.

15. When finding the fifth roots of 1, we expressed the nonreal roots only in polar form. Here we'll express them exactly in radical form.

    (a) Draw regular pentagon $ABCDE$ with each side of length 1. Each angle of the pentagon is $180°$. Now draw diagonals AC and AD. Determine all the angles in the diagram.

    (b) Let $AC = AD = x$. Use the Law of Cosines on triangle AED to show that $x^2 = 2 - 2\cos(108°) = 2 + 2\cos(72°)$.

    (c) Use the Law of Cosines on triangle ACD to show that $\cos(72°) = \frac{1}{2x}$.

    (d) Substitute the result of (c) into the equation from (b) to obtain the cubic equation $x^3 - 2x - 1 = 0$.

    (e) This is a factorable equation. Solve it for x. Which of the roots makes sense for this problem?

    (f) Show that $\cos(72°) = \dfrac{\sqrt{5} - 1}{4}$.

    (g) Show that $\sin(72°) = \dfrac{\sqrt{10 + 2\sqrt{5}}}{4}$.

Similar calculations will show that $\cos(144°) = \dfrac{-\sqrt{5} - 1}{4}$ and $\sin(144°) = \dfrac{\sqrt{10 - 2\sqrt{5}}}{4}$.

    (h) Express all the fifth roots of 1 in radical form.

16. Prove that $\sqrt{2} + \sqrt[3]{3}$ is irrational. [*Hint*: Show that $\sqrt{2} + \sqrt[3]{3}$ is a root of $x^6 - 6x^4 - 6x^3 + 12x^2 - 36x + 1 = 0$.]

17. Let p be a polynomial with integer coefficients and let r_1, r_2, r_3, r_4 be *distinct* integers such that $p(r_1) = p(r_2) = p(r_3) = p(r_4) = 7$.

(a) Show that $p(x) - 7 = (x - r_1)(x - r_2)(x - r_3)(x - r_4)q(x)$, where q is a polynomial.

(b) Show that there is no integer value of x such that $p(x) = 14$.

18. Let $p(x) = ax^3 + bx^2 + cx + d$ be a polynomial with integer coefficients. Show that if $p(0)$ and $p(1)$ are odd, then the equation $p(x) = 0$ has no integer solutions.

Chapter 4
Combinatorics and Graph Theory

There are two related branches of mathematics that are especially conducive to experimentation and conjecture. Combinatorics is the mathematics of counting large numbers efficiently (such as the number of distinct poker hands or the number of ways of choosing the numbers in a lottery game). Graph theory (which has nothing to do with the kind of graphs you draw for functions) is the study of mathematical objects consisting of a set of vertices with line segments connecting some, or all, pairs of vertices.

Both combinatorics and graph theory are extremely useful in many applied mathematical problems. For example, combinatorics plays an important role in probability and statistics. Graph theory has many uses in computer science and operations research.

We could devote an entire course to these topics (and such courses, often called Discrete Mathematics, do exist in many schools). However, we'll settle for just a taste of the sort of problems that arise in these areas.

4.1 Combinations and the Binomial Theorem

We'll begin this section with an investigation of a certain family of polynomials. Let $p_n(x) = (1 + x)^n$, where n is a nonnegative integer. Clearly, by expansion, we can write $p_n(x)$ in polynomial form.

1. Expand $p_n(x)$ for $n = 2, 3, 4$.

2. How many terms are there in the expansion of $p_n(x)$? What is the first term? What is the last term?

As it stands now, we have to do considerable algebraic manipulations to expand $p_n(x)$ as a polynomial. For the most part, we have to do it recursively; that is, if we want $p_6(x)$, we have to compute $p_n(x)$ for at least some values of n less than 6. We'd like to devise a method for computing $p_n(x)$ directly for any given value of n.

Certainly, $p_n(x) = a_0 + a_1x + a_2x^2 + \cdots + a_nx^n$. Our task is to determine the coefficients $a_0, a_1, \ldots, a_n$. We need to introduce notation that reflects the fact that the coefficients depend on both n and the exponent of the term to which they belong. Let $\binom{n}{k}$ represent the coefficient of x^k in the expansion of $p_n(x) = (1 + x)^n$. In other words,

$$p_n(x) = (1 + x)^n = \binom{n}{0} + \binom{n}{1}x + \binom{n}{2}x^2 + \cdots + \binom{n}{n}x^n = \sum_{k=0}^{n} \binom{n}{k}x^k.$$

The notation $\binom{n}{k}$ is read "n choose k" for reasons that we shall see later. These numbers are also called *binomial coefficients* since they arise from expanding the binomial $(1 + x)^n$.

We seek a convenient method of computing $\binom{n}{k}$, for any n and any $0 \leq k \leq n$.

3. *Determine* $\binom{3}{1}$, $\binom{3}{2}$, $\binom{4}{1}$, $\binom{4}{2}$, *and* $\binom{4}{3}$ *by direct computation.*

4. *Determine* $\binom{n}{0}$, $\binom{n}{1}$, *and* $\binom{n}{n}$ *for all n.*

First we'll develop a recursive relationship that $\binom{n}{k}$ must satisfy. Note that

$$p_{n+1}(x) = (1 + x)p_n(x) = p_n(x) + xp_n(x).$$

5. *By equating coefficients of x^k on both sides of this equation, prove the following theorem.*

Theorem 4.1 $\binom{n + 1}{k} = \binom{n}{k} + \binom{n}{k - 1}$ for all positive integers n and all $k = 1, 2, 3, \ldots, n$.

6. *Use Theorem 4.1 to compute* $\binom{5}{2}$, $\binom{5}{3}$, *and* $\binom{6}{3}$.

Theorem 4.1 can be used to create an array for storing the binomial coefficients. The array is called *Pascal's triangle*.[1] The first few rows are given below.

$$
\begin{array}{ccccccc}
1 & & & & & & \\
1 & 1 & & & & & \\
1 & 2 & 1 & & & & \\
1 & 3 & 3 & 1 & & & \\
1 & 4 & 6 & 4 & 1 & & \\
1 & 5 & 10 & 10 & 5 & 1 &
\end{array}
$$

If we start numbering the rows with $n = 0$ and start numbering the columns with $k = 0$, then the entry in row n, column k is $\binom{n}{k}$. Notice that Theorem 4.1 says that each entry is the sum of the entry above it and the entry to the left of the one above it.

7. *Compute the next three rows of Pascal's triangle.*

8. *Find* $\binom{7}{3}$, $\binom{7}{5}$, *and* $\binom{8}{4}$ *in Pascal's triangle.*

9. *Complete and prove the following theorems.*

Theorem 4.2 $\displaystyle\sum_{k=0}^{n}\binom{n}{k} = \underline{\hspace{3cm}}$, for $n \geq 0$.

Theorem 4.3 $\displaystyle\sum_{k=0}^{n}(-1)^k\binom{n}{k} = \underline{\hspace{3cm}}$, for $n \geq 1$.

Note: Theorem 4.3 is not true for $n = 0$. (Why?)

We still have not devised a method of computing the binomial coefficients directly; Theorem 4.1 provides only a recursive scheme. Thus, computing $\binom{14}{8}$ still requires us to compute at least parts of the first 13 rows of Pascal's triangle.

---

[1] Blaise Pascal (1623–1662) was a French mathematician who made many contributions to geometry and other areas, beginning at age 12. Arguably, he could have been one of the greatest mathematicians of all time had he not suffered from ill health and periodically given up mathematics to study religion. The array known as Pascal's triangle was described in 1653, although there are references to it in Chinese mathematics as early as 1303. Pascal is credited with discovery many of the properties of the triangle, such as those we are investigating here.

We do have direct (trivial) means of calculating $\binom{n}{k}$ for $k = 0$ and $k = 1$.

Let's see what happens for $k = 2$. We know that $\binom{2}{2} = 1$. Observe that $\binom{3}{2}$

$= \binom{2}{2} + \binom{2}{1} = 1 + 2 = 3$, $\binom{4}{2} = \binom{3}{2} + \binom{3}{1} = 3 + 3 = 6$, and so on.

In general, Theorem 4.1 tells us that $\binom{n}{2} = \binom{n-1}{2} + n - 1$.

10. *Argue that* $\binom{n}{2} = \sum_{j=1}^{n-1} j = \dfrac{n(n-1)}{2}$, *for $n \geq 2$.*

11. *Suggest and prove a formula for* $\binom{n}{3}$. *[Hint: You might find it useful to express the ratio of* $\binom{n}{3}$ *to* $\binom{n}{2}$ *in terms of n.]*

12. *Prove the following theorem.*

Theorem 4.4 $\binom{n}{k} = \dfrac{n!}{k!(n-k)!}$, for $n \geq 0$, $0 \leq k \leq n$.

Note: Most calculators can compute $\binom{n}{k}$ for any n and k. Sometimes the button may be labeled $C(n, r)$ or $_nC_r$, or it may be listed under a "probability" or "combinatorics" menu.

Observe that there is a certain symmetry in the binomial coefficients. This is reflected in the fact that each row of Pascal's triangle reads the same left-to-right as right-to-left.

13. *Complete and prove the following corollary.*

Corollary 4.1 $\binom{n}{k} = \binom{n}{??}$.

Although we have so far just considered the polynomial $p_n(x) = (1 + x)^n$, the results we've developed can be adapted to the expansion of any binomial of the form $(a + b)^n$. The result is called the *Binomial Theorem*.

Theorem 4.5 (Binomial Theorem) $(a + b)^n = \sum_{k=0}^{n} \binom{n}{k} a^{n-k} b^k$, where

$\binom{n}{k}$ is given by Theorem 4.4.

The proof of Theorem 4.5 is by induction—see question 2 at the end of the chapter.

14. *Expand each of the following.*

    (i) $(3 + x^2)^4$ *(ii)* $\left(y + \dfrac{1}{y} \right)^6$

15. *Determine the coefficient of z^9 in the expansion of $(2 - z^3)^8$.*

4.1.1 Combinations

The binomial coefficients can be approached from an entirely different point of view, one that involves a branch of mathematics called *combinatorics*. Combinatorics deals with determining efficient methods for counting large numbers. For example, suppose we wanted to determine the number of different five-card poker hands. Making a list of all the possibilities would be quite a tedious process as there are more than two million different hands. However, a few simple combinatorial formulas reduces the difficulty of this problem immensely.

We begin with some terminology. Recall that a *set* is a well-defined collection of distinct objects or *elements*. The order in which the elements are listed is irrelevant, so $\{a, b, c\}$ is the same as $\{b, c, a\}$. We say that a set T is a *subset* of S, denoted $T \subseteq S$, if every element of T is also an element of S. The problem of counting poker hands is equivalent to counting the number of 5-element subsets of a set of 52 elements (the deck of cards).

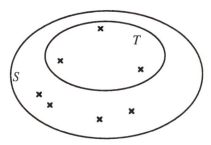

More generally, we will be concerned with counting the number of k-element subsets of a set with n elements. We will show the following.

Theorem 4.6   The number of k-element subsets of a set with n elements is the binomial coefficient $\binom{n}{k}$.

In this context, the number $\binom{n}{k}$ is often defined as the "number of *combinations* of n objects taken k at a time." The word *combination* means unordered selection of objects.

Before we actually prove Theorem 4.6, let's gather some supporting evidence. First note that S itself is considered to be a subset of S, and the *empty set* (or *null set*), denoted $\varnothing$, is also a subset of S. In other words, there is one subset with 0 elements and one subset with n elements. This agrees with the fact that

$$\binom{n}{0} = \binom{n}{n} = 1.$$

16. *Make up a set with four elements and count all the subsets of sizes 1, 2, and 3. Do the answers agree with the corresponding binomial coefficients? What is the total number of subsets?*

17. *Make up a set with five elements and count all the subsets of sizes 1, 2, 3, and 4. Do the answers agree with the corresponding binomial coefficients? What is the total number of subsets?*

18. *Interpret and justify Corollary 4.1 in the context of subsets. [Hint: Every time we pick a subset of size k, what else do we get?]*

Now let's look at the interpretation of Theorem 4.1 in the context of subsets. Let $S = \{a_1, a_2, a_3, \ldots, a_n, a_{n+1}\}$ be a set with $n + 1$ elements. If Theorem 4.6 is true, then the number of k-element subsets of S is $\binom{n + 1}{k}$. These subsets can be partitioned into two groups—those that contain a_1 and those that don't.

19. *How many subsets contain a_1? How many do not contain a_1? Does this support Theorem 4.1?*

20. *Theorem 4.2 makes a claim about the total number of subsets of a set with n elements. What is this claim? Prove it by induction.*

Although we have gathered a lot of evidence in support of Theorem 4.6, we have not actually proven it. We do so now.

Consider the product $(1 + x_1)(1 + x_2) = 1 + x_1 + x_2 + x_1x_2$. Notice that there is a term corresponding to every possible subset of $\{x_1, x_2\}$, where we equate the "1" with the null set. Moreover, if we now set $x_1 = x_2 = x$, we get $(1 + x)^2 = 1 + 2x + x^2$. The coefficient of x is precisely the number of 1-

element subsets of $\{x_1, x_2\}$; the coefficient of x^2 is the number of 2-element subsets.

This argument easily generalizes. The expansion of $(1 + x_1)(1 + x_2) \cdots (1 + x_n)$ contains a term corresponding to each subset of $\{x_1, x_2, \ldots, x_n\}$. Upon setting $x_1 = x_2 = \cdots = x_n = x$, we see that the coefficient of x^k in the expansion of $(1 + x)^n$ is precisely the number of k-element subsets of $\{x_1, x_2, \ldots, x_n\}$. By definition, that coefficient is $\binom{n}{k}$. Hence, the proof of Theorem 4.6 is complete.

21. *How many different 4-person committees can be made from a group of 10 people? Assume that the order in which the people are placed in the committee is irrelevant.*

22. *How many different 5-card poker hands can be dealt from a standard deck of 52 cards? Assume that the order in which the cards are dealt is irrelevant.*

Keep in mind that these results apply only when the selection of objects is unordered. So, for example, it would not apply if we wanted to count the number of ways of picking the top 3 finishers in order in a horse race with 10 horses.

4.1.2. Some Combinatorial Identities

There are a number of interesting facts about the binomial coefficients that are worth exploring. We've already seen some in Theorems 4.1, 4.2, and 4.3. For the first two of these, we gave two arguments—one involving the basic definition of the binomial coefficients from the expansion of

$$p_n(x) = (1 + x)^n = \sum_{k=0}^{n} \binom{n}{k} x^k$$

and one involving the counting of subsets.

Now consider the sum

$$S_n = \sum_{k=0}^{n} k \binom{n}{k}.$$

23. *Hypothesize a formula for S_n.*

24. *Prove your hypothesis in two different algebraic ways:*

(i) *by using the fact that* $\binom{n}{k} = \dfrac{n!}{k!(n - k)!}$.

(ii) *by computing* $p_n'(x)$.

It is also possible to prove this result by combinatorial methods. This approach involves finding an appropriate counting problem and solving it by two different schemes. One scheme will give you the left-hand side of the formula you wish to prove; the other will give you the right-hand side. It takes a lot of experience— more than you would have gotten by reading this book so far—to think of the right counting problem. However, you should be able to fill in the details of the proof sketched below.

Imagine a group of n people from which you are to select a committee of size $k \geq 1$. Then you appoint a chair of the committee. The problem is to determine the total number of committees with appointed chair. This is different from merely counting the number of committees since two committees with the same people but a different chair are considered different. To illustrate, let's make a list of all the possibilities for the case $n = 3$. The chair is underlined.

$k = 1$: $\{\underline{a}\}$, $\{\underline{b}\}$, $\{\underline{c}\}$
$k = 2$: $\{\underline{a}, b\}$, $\{\underline{b}, a\}$, $\{\underline{a}, c\}$, $\{\underline{c}, a\}$, $\{\underline{b}, c\}$, $\{\underline{c}, b\}$
$k = 3$: $\{\underline{a}, b, c\}$, $\{\underline{b}, a, c\}$, $\{\underline{c}, a, b\}$

So there are 12 such committees. Notice that for $k = 2$, we get twice as many committees as we would have had if we had not appointed a chair. For $k = 3$, we get three times as many.

25. *Argue that, in general, the total number of committees is*

$$S_n = \sum_{k=0}^{n} k \binom{n}{k}.$$

We could have counted the committees differently. First let's appoint the chair; then, we'll fill up the rest of the committee.

26. *How many choices are there for the chair?*

27. *How many different committees (of any size ≥ 1) can be formed for each possible choices of chair?*

28. *What is the total number of committees with chair? Does this prove your hypothesis from question 23?*

Now consider the sum

$$S_n = \sum_{k=0}^{n} \binom{n}{k}^2.$$

29. *Hypothesize a formula for S_n. [Hint: The answer is one of the binomial coefficients.]*

A direct algebraic proof using Theorem 4.4 is difficult. Consequently, we'll try two other proofs—one using the Binomial Theorem and one involving a combinatorial argument.

For the first proof, consider the expression $(1 + x)^{2n} = (1 + x)^n(1 + x)^n$.

30. What is the coefficient of x^n in the binomial expansion of the left side?

If we expand the two factors on the right side and multiply them together, we get

$$\left[\binom{n}{0} + \binom{n}{1}x + \binom{n}{2}x^2 + \cdots + \binom{n}{n}x^n\right]\left[\binom{n}{0} + \binom{n}{1}x + \binom{n}{2}x^2 + \cdots + \binom{n}{n}x^n\right].$$

We need to find the coefficient of x^n in this expression. It may be difficult to see what's going on in general, so we'll try a special case first.

Let $n = 2$. So we are looking for the coefficient of x^2 in the expansion of $(1 + 2x + x^2)(1 + 2x + x^2)$.

31. What is that coefficient?

Note that there are three terms contributing to that coefficient: $1(1) + 2(2) + 1(1)$.

32. More generally, argue that the coefficient of x^n the expansion of

$$\left[\binom{n}{0} + \binom{n}{1}x + \binom{n}{2}x^2 + \cdots + \binom{n}{n}x^n\right]\left[\binom{n}{0} + \binom{n}{1}x + \binom{n}{2}x^2 + \cdots + \binom{n}{n}x^n\right]$$

is

$$\binom{n}{0}\binom{n}{n} + \binom{n}{1}\binom{n}{n-1} + \binom{n}{2}\binom{n}{n-2} + \cdots + \binom{n}{n}\binom{n}{0}.$$

33. Show that this coefficient can be written as

$$S_n = \sum_{k=0}^{n} \binom{n}{k}^2.$$

For the combinatorial argument, assume we have a group of n men and n women. We wish to select a committee of size n.

34. In how many ways can we do this?

35. How many of the committees have no men?

36. How many have 1 man and $n - 1$ women? 2 men and $n - 2$ women?

37. Argue that the total number of committees is

$$\binom{n}{0}\binom{n}{n} + \binom{n}{1}\binom{n}{n-1} + \binom{n}{2}\binom{n}{n-2} + \cdots + \binom{n}{n}\binom{n}{0}$$

which, by the argument used above, is equivalent to the desired summation.

4.2.   Generating Functions

In the previous section, we considered the problem of counting the number of subsets of size k that could be formed from a set of n objects. Implicit in this discussion is the requirement that the n objects be distinct. In this section, we'll allow for the possibility that some of the objects are repeated.

To get started, let's look more closely at the problem in which the objects are distinct. Suppose we have three objects, $\{a, b, c\}$. The number of ways of picking zero objects is $\binom{3}{0}$. The number of ways of picking one object is $\binom{3}{1}$. The number of ways of picking two objects is $\binom{3}{2}$, and the number of ways of picking three objects is $\binom{3}{3}$. These numbers appear as coefficients in the binomial expansion of $(1 + x)^3 = 1 + 3x + 3x^2 + x^3$.

In general, if we have n distinct objects, the number of ways of picking a subset of size k is $\binom{n}{k}$, which is the coefficient of x^k in the expansion of $(1 + x)^n$. So, as we have already seen, there appears to be a close connection between polynomials (of which the binomial expansions are a special case) and counting problems.

This suggests that we might approach more complex counting problems by creating a polynomial (or, in some cases, a power series with infinitely many terms) in which the coefficient of x^k represents the number of "subsets" of size k.

Suppose we have a box with two indistinguishable green marbles and two indistinguishable red marbles.

1. In how many ways can you pick one marble? two marbles? three marbles? four marbles? List the possible outcomes for each case.

2. Create a fourth-degree polynomial in which the coefficient of x^k is the number of ways of picking k marbles, $k = 0, 1, 2, 3, 4$. (We'll agree that there *is one way of picking no marbles; hence, the constant term in the polynomial is 1.)*

3. Show that your answer to question 2 can be factored as $(1 + x + x^2)^2$.

4. Repeat questions 1 and 2 if the box contains two green, one red, and one blue marble.

5. Show that the polynomial you found in question 4 can be factored as $(1 + x + x^2)(1 + x)^2$.

Let $a_0, a_1, \ldots, a_n$ be a sequence of numbers. The polynomial $p(x) = \sum_{k=0}^{n} a_k x^k$ is called a *generating function* for the sequence. Generating functions are a useful tool in many areas of mathematics. Here, we'll see how they are applied to some combinatorial problems.

There is a one-to-one correspondence between sequences of numbers and generating functions. In other words, every sequence of numbers has a unique generating function associated with it and, more importantly, every generating function arises from a unique sequence.

In this section, we will let a_k denote the number of ways of picking k objects. Without actually counting, we'll create the generating function. Then we can recover the values of a_k by looking at the coefficients of the generating function.

Let's return to the simple problem in which we have three distinct objects $\{a, b, c\}$. Any subset will have: "either no a's **or** 1 a" **and** "either no b's **or** 1 b" **and** "either no c's **or** 1 c." The corresponding generating function is $(1 + x)^3 = (1 + x)(1 + x)(1 + x)$. The first factor of $1 + x$ corresponds to the first object and results from the fact that there is one way to pick no a's and one way to pick 1 a. The variable x is just a placeholder; we are really interested in the coefficients. The other two factors of $1 + x$ correspond to objects b and c. The factors are multiplied because we need to account for the absence or presence of all objects simultaneously.

Now let's look at the problem in which there are two green and two red marbles. Since there is one way to pick no greens, one way to pick one green (because they are indistinguishable), and one way to pick two greens, then the generating function for the green marbles alone is $1 + x + x^2$. Using the same logic as above, the generating function for the whole problem is $(1 + x + x^2)^2$.

6. Justify the generating function in question 5.

7. Suppose the box has three green marbles, two red marbles, and one blue marble. Create the generating function for this problem.

8. Use the generating function to determine the number of ways of picking four marbles.

9. A fruit bowl contains two apples and one piece each of n − 2 other fruits. Use a generating function to show that the number of ways of picking k pieces of fruit is

$$\binom{n-2}{k} + \binom{n-2}{k-1} + \binom{n-2}{k-2}.$$

10. Give a more direct combinatorial argument of the result of question 9.

11. There are three households. One of them has two people living in it, while the other two have one person each. In how many ways can a team of surveyors select k people if it agrees to take either all the people in a given household or none of the people in that household?

Now let's suppose that there is an infinite supply of red and green marbles. The generating function for the red and green marbles separately is $1 + x + x^2 + \cdots$. Consequently, the generating function for the whole problem is $(1 + x + x^2 + \cdots)^2$.

12. Show that the expression above is equivalent to $\dfrac{1}{(1-x)^2}$.

At this point, we can proceed in two directions. In one, we use an extension of the binomial theorem for negative integer exponents. To do so, we have to define the binomial coefficents $\binom{n}{k}$ for values of n that are not positive integers. This is not hard if we remember from Theorem 4.4 that one formula for computing $\binom{n}{k}$ is

$$\frac{n(n-1)(n-2) \cdots (n-k+1)}{k!},$$

where there are k factors in the numerator. This makes sense for all values of n, as long as k is a positive integer.

13. Compute $\binom{-4}{3}$, $\binom{1/2}{2}$, and $\binom{3/2}{4}$.

14. Show that $\dfrac{1}{(1-x)^2} = 1 + 2x + 3x^2 + 4x^3 + \cdots$.

15. In how many ways can we pick k marbles? Why does this make sense?

Another approach to this problem hinges on the observation that the process of expanding a function as a polynomial is equivalent to writing a Taylor series for the function. Recall that the Taylor series (centered at $x = 0$) for a function f is of the form

$$f(0) + f'(0)x + \frac{f''(0)}{2!}x^2 + \frac{f'''(0)}{3!}x^3 + \cdots.$$

It follows that the coefficient

$$a_k = \frac{f^{(k)}(0)}{k!}.$$

16. *Verify that the Taylor series for* $\dfrac{1}{(1-x)^2}$ *is the same as the answer to question 14.*

17. *Extend these results to the case in which there is an infinite supply of three different colors of marbles.*

There are many variations of this theme that we could pursue. Suppose that we had two colors of marbles in infinite supply but we must pick an even number of each color.

18. *Write a generating function for this problem and use it to determine the number of ways of picking k marbles. (Clearly, the answer is 0 if k is odd.)*

4.3. Introduction to Graph Theory

The old village of Königsburg in Germany was situated on the Pregel River. There were two islands in the middle of the river. The islands were connected to each other and to the river banks by seven bridges, as shown in the diagram below. Villagers would amuse themselves by trying to start on one side of the river, cross each bridge exactly once, and return to their starting point.

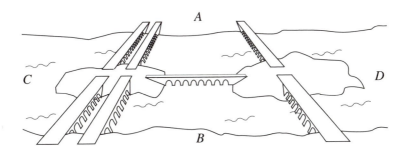

1. *Is this possible?*

2. If this is not possible, what is the greatest number of bridges you can cross without retracing any while returning to your starting point? What if you don't have to return to the starting point?

3. If you could build another bridge, would that help and, if so, where would you build it? Consider the cases in which you have to return to the starting point and in which you don't.

Now consider each of the diagrams below.

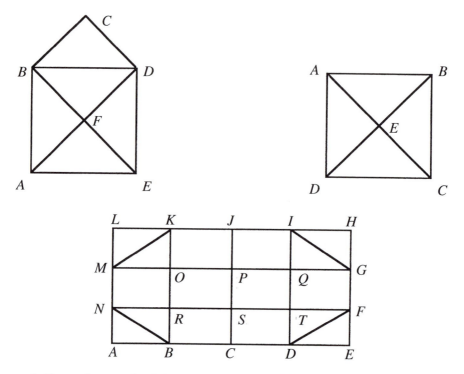

4. Try to draw each of these figures without lifting your pencil and without retracing any line segments.

Before we proceed, we need a few definitions.

Definition   A *graph* consists of a set of points, called the *vertices*, and a set of line segments, called the *edges*, that connect the vertices.

Each of the diagrams above is a graph. In the first diagram, there are six vertices and ten edges. In the second diagram, there are five vertices and eight edges.

5. *How many vertices and edges are there in the third diagram?*

The only thing that is important in a graph is the number of vertices, the number of edges, and the way in which the vertices are connected. The actual distances and orientation of the graph do not matter. For example, the two graphs below are equivalent.

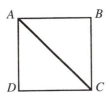 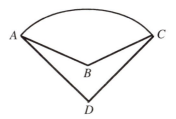

Note: Another word for "equivalent" in this context is *isomorphic*. This is a word that has meanings in many different branches of mathematics. Determining whether two graphs are isomorphic is actually a very difficult problem, and we will not pursue it here.

The Königsburg bridge problem can now be restated in terms of graphs. The graph will have four vertices (one for each river bank and island) and seven edges (one for each bridge). We want to find a path that uses every edge exactly once.

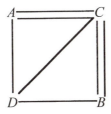

Definition   An *Euler path*[2] is a sequence of vertices that traces each edge exactly once. An *Euler cycle* is an Euler path that begins and ends at the same vertex.

Clearly, some graphs have Euler paths and some do not. So, question 4 really asks whether each graph has an Euler path. More specifically, there should be

---

[2] Leonhard Euler (1707–1783) was a Swiss mathematician who made important contributions to many areas of mathematics. We have already seen his name in the context of the Euler-Fermat Theorem in Chapter 2. It was Euler who first solved the Königsburg bridge problem in 1736.

some relationship between the number of edges and the number of vertices that indicates whether the graph has an Euler path.

To determine this relationship, let's begin with a few simple examples. Certainly, a simple square has an Euler path and we can begin the path at any vertex.

> 6. *Now draw a square with one diagonal. Is there an Euler path? If so, at which vertex (or vertices) does it begin?*

> 7. *What happens if you draw both diagonals?*

Note that if you draw the second diagonal in the normal way, you get another vertex at the point where the diagonals cross. Although this does not change the answer to question 7, you might want to draw the second diagonal around the outside, as we showed in the diagram following question 5.

It seems plausible that the existence of an Euler path depends on the number of edges meeting at each vertex. We call this number the *degree* of the vertex.

> 8. *For each graph considered so far, determine the degree of each vertex.*

> 9. *What must be true about the sum of the degrees of all the vertices in a graph?*

> 10. *Prove that there is always an even number of vertices of odd degree.*

> 11. *Complete the following theorem.*

Theorem 4.7

(i) A graph G has an Euler path if and only if _____ _____. The Euler path begins at _____ and ends at _____.

(ii) G has an Euler cycle if and only if _____.

Notice that this theorem provides both a necessary and sufficient criterion for determining whether a graph has an Euler path (or cycle). The necessity of this criterion is fairly obvious; its sufficiency is less so. In other words, our experimentation above is convincing that any graph that has an Euler path must satisfy the criterion (or, equivalently, any graph that doesn't satisfy the criterion can't have an Euler path). It takes a bit more work to show that any graph that satisfies the criterion has an Euler path. We won't pursue this now.

4.3.1. Planar Graphs and Euler's Formula

Consider the following problem: Each of three houses is to be connected to three utilities: water, gas, and electricity. It is desired to make these connections in such a way that none of the connections intersect.

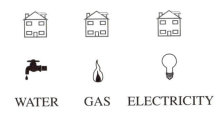

WATER   GAS   ELECTRICITY

12. Is this possible?

Definition   If a graph can be drawn so that none of the edges intersect except at the vertices, then the graph is said to be *planar.*

13. Which of the following graphs are planar?

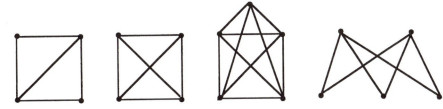

Note: Determining whether a graph is planar is not an easy task. There is a theorem, called *Kuratowski's Theorem*, that gives a criterion for a graph to be planar. Although the criterion is easy to state, it is not so easy to implement.[3] We won't pursue this here.

Every planar graph creates a number of regions, or *faces*. In the first graph above, there are three faces—the two triangles and the region outside the square.

14. Draw several planar graphs. For each one, count the number of vertices, the number of edges, and the number of faces.

By counting the "exterior" region as a face, we are, in essence, imagining that we are drawing the graph on the surface of a sphere. Then, for instance, a triangular graph really does divide the surface of the sphere into two (finite) pieces.

---

[3] For a discussion of Kuratowski's Theorem, see *Graph Theory* by Frank Harary, Addison-Wesley, 1969, or any of a number of other graph theory texts.

Suppose we started with a simple triangular graph. It has three edges, three vertices, and two faces. We could make the graph more complicated in several ways:

- Add a vertex to one side.
- Add a vertex not on one of the edges and connect it to one of the existing vertices with an edge.
- Add a triangular face that shares an edge with the existing triangle.

15. What effect does each of these have on the number of vertices, number of edges, and number of faces?

There is a relationship, known as *Euler's formula*, between the number of vertices v, the number of edges e, and the number of faces f. It is valid for all planar graphs, whether we draw them on a flat piece of paper (in which case we count the exterior region as a face) or on the surface of a sphere.

16. Complete the following theorem.

Theorem 4.8   Let G be a planar graph. Then ———————— .

To prove Euler's formula, we'll use mathematical induction. There are three variables—v, e, and f—that could serve as the induction variable. It turns out to be easiest to use f.

Let G be a graph with f faces. Assume that Euler's formula holds for G. We need to show that it holds for a graph with $f + 1$ faces.

Let's add another face to G, creating a new graph G'. The face could be a "triangle," "quadrilateral," "pentagon," and so on. (We're using these terms loosely, since the edges don't have to be straight line segments.)

17. If the face is a "polygon" with n vertices, how many vertices did we add to G? How many edges did we add?

18. Show that Euler's formula holds for G'.

Euler's formula can be used to prove an interesting result in solid geometry. A *polyhedron* is a solid whose surface consists of a number of polygonal faces. The polyhedron is said to be *regular* if all the faces are congruent and all the angles at the vertices are equal. For example, a cube is a regular polyhedron. We will show that there are only five regular polyhedra. Note that any polyhedron, regular or not, can be distorted into a graph on the surface of a sphere and, thus, is equivalent to a planar graph. Hence, Euler's formula holds.

One property of a regular polyhedron is that each face has the same number of edges and each vertex has the same number of edges meeting there. (Think about a cube: There are four edges on each face and three edges meeting at each vertex. On the other hand, a pyramid with a square base, which is not regular, has four faces with three edges and one face with four edges.)

Let p = the number of edges on each face and q = the number of edges at each vertex.

19. Argue that pf = *2e and qv* = *2e.*

Upon solving these for f and v and substituting in Euler's formula, we get

$$\frac{2e}{q} - e + \frac{2e}{p} = 2.$$

20. Show that this implies $\dfrac{1}{q} + \dfrac{1}{p} = \dfrac{1}{2} + \dfrac{1}{e} > \dfrac{1}{2}.$

It is clear that $p \geq 3$ and $q \geq 3$. Otherwise we wouldn't get a three-dimensional figure.

21. Show that there are only five solutions to the equation in question 20.

The five regular solids are the tetrahedron, cube, octahedron, dodecahedron, and icosahedron.

4.4. Additional Questions

1. Use the binomial theorem to derive formulas for each of the following.

(a) $\displaystyle\sum_{k=0}^{n} k^2 \binom{n}{k}$

(b) $\displaystyle\sum_{k=0}^{n} \frac{1}{k+1} \binom{n}{k}$

[*Hint:* Think calculus. For (a), note that $k^2 = k(k-1) + k$.]

2. Prove Theorem 4.5.

3. Let x and y be real numbers such that $x + \frac{1}{x} = y$. Express each of the following in terms of y.

(a) $x^2 + \dfrac{1}{x^2}$ (b) $x^3 + \dfrac{1}{x^3}$ (c) $x^4 + \dfrac{1}{x^4}$

4. Prove $\left[\displaystyle\sum_{k=0}^{n} \binom{n}{k}\right]^2 = \displaystyle\sum_{k=0}^{2n} \binom{2n}{k}.$

5. Let S be a set with n elements. What is the total number of subsets of S that have an odd number of elements? Prove your result.

6. Hypothesize and prove a formula for $\dbinom{n+2}{3} - \dbinom{n}{3}$, for $n \geq 3$.

7. Prove that $\sum_{k=0}^{n-1}\binom{n}{k}\binom{n}{k+1} = \binom{2n}{n+1}$. [*Hint:* Consider the expansion of $(1 + x)^{2n}$.]

8. (Requires calculus.) Show that the expansion of $p_n(x) = (1 + x)^n$ obtained by the Binomial Theorem is the same as the one obtained by writing the Taylor series for $p_n(x)$.

9. Use the Binomial Theorem to write the first four terms in the expansion of $(1 + x)^{1/2}$. Then use your result to estimate $\sqrt{2}$.

10. Which is bigger: 10^n or $11^n - 9^n$? Prove your result.

[*Hint:* $11 = 10 + 1, 9 = 10 - 1$.]

11. An ice-cream parlor has 10 flavors. You wish to create a sundae with three scoops. In how many ways can you do this if:
    (a) you use three different flavors?
    (b) you may use more than one scoop of a flavor?
    (c) Express your answer to part (b) as a single binomial coefficient.

(Assume that the order in which the flavors are used is irrelevant—for example, {chocolate, vanilla, pistachio} is the same as {pistachio, chocolate, vanilla}.)

12. When creating a subset, we agree that the order of the elements within the subset is irrelevant. Thus, $\{a, b\} = \{b, a\}$. Now let's assume that order *is* relevant, so that $\{a, b\}$ and $\{b, a\}$ are different. To avoid confusion, we'll use the term *arrangement* rather than *subset*.
    (a) Write all the 2-element arrangements from the set $\{a, b, c, d\}$.
    (b) Repeat (a) for 3-element arrangements.
    (c) In general, how many k-element arrangements can be obtained from each k-element subset?
    (d) Derive a formula for the number of k-element arrangements that can be formed from a set of n elements.
    (e) How many different signals can be made from a set of eight different colored flags if a signal consists of three flags placed in order on a flagpole?

13. Suppose there is an infinite supply of each of n different colored marbles. Determine the generating function for the entire problem. In how many different ways can we pick k marbles?

14. Suppose you have an infinite supply of $1 and $2 bills. (Wishful thinking!) The goal of this problem is to determine the number of different ways in which you can pay for an item that costs k dollars. For example, there are three ways you can pay for a $5 item—five $1 bills, two $2 bills, and one $1 bill, or one

$2 bill and three $1 bills. To derive the generating function for this problem, first assume that you use only $1 bills. There is exactly one way in which to pay k dollars, for every k. Thus, the generating function for the $1 bills is $1 + x + x^2 + x^3 + \cdots$.

(a) Determine the generating function for $2 bills. [*Hint*: How many ways can you pay for a $3 item with $2 bills only? a $4 item?]

(b) Show that the generating function for the entire problem is

$$p(x) = \frac{1}{(1 - x)(1 - x^2)}.$$

(c) The next step is to expand $p(x)$ as a polynomial. We could multiply the two infinite series term by term, or we could use a Taylor series approach. Since both methods are a bit tedious, we'll try something different. Show that

$$p(x) = \frac{1 + x}{(1 - x^2)^2} = (1 + x)\frac{1}{(1 - x^2)^2}.$$

(d) In question 14 of Section 4.2, we showed that

$$\frac{1}{(1 - x)^2} = 1 + 2x + 3x^2 + \cdots.$$

Modify this to get the polynomial expansion of $\dfrac{1}{(1 - x^2)^2}$.

(e) Use your answer to part (d) to determine the first 10 terms of the polynomial expansion of $p(x)$.

(f) In how many ways can you pay for an item costing k dollars?

15. Now suppose that you have one $1 bill, one $2 bill, one $4 bill, one $8 bill, and one $16 bill. Again, our goal is to determine the number of ways in which you can pay for an item costing k dollars.

(a) Argue that the generating function for this problem is

$$p(x) = (1 + x)(1 + x^2)(1 + x^4)(1 + x^8)(1 + x^{16}).$$

(b) Show that $p(x) = \dfrac{1 - x^{32}}{1 - x}$.

(c) Expand the expression in part (b) as a polynomial. [*Hint:* Think about geometric series.]

(d) Interpret the result from part (c).

16. The generating function for a certain counting problem is $p(x) = \dfrac{1-2x}{1-3x}$. Determine the coefficient of x^k.

17. What is the maximum number of edges in a graph with n vertices if there is no more than one edge connecting any pair of vertices?

18. A graph consists of a checkerboard with n squares on each side. Show that Euler's formula holds for this graph.

Chapter 5
Difference Equations and Iteration

In this chapter, we'll begin to tie together some of the ideas of previous chapters. Specifically, we'll investigate what happens when a function (which may or may not be a polynomial) is applied repeatedly to a particular input. In other words, we'll consider the sequence $\{x_0, f(x_0), f(f(x_0)), f(f(f(x_0))), \ldots\}$. Depending on the nature of f and the value of x_0, the sequence may converge to a finite value, may bounce back and forth between two or more values, or may become infinitely large. Our goal is to determine which, if any, of these behaviors occurs.

Before we can do this, we need to take a closer look at sequences, with particular emphasis on the conversion between explicit and recursive definitions. (The sequence above is defined recursively.) This involves a brief discussion of difference equations, a topic that has many applications in a variety of areas, not the least of which is combinatorics. Unfortunately, a detailed discussion of difference equations would divert us from our main objective, so we'll forgo such a discussion at this time. You can, however, read more about them in any book on discrete mathematics, combinatorics, or dynamical systems.

5.1. Linear Difference Equations

In Chapter 1, we studied sequences of numbers $\{u_1, u_2, u_3, \ldots\}$. We defined the sequence in one of two ways. One is explicit; that is, we give a formula for u_n in terms of n. The other method is recursive; that is, we express u_n as a function of one or more previous terms in the sequence. In some cases, such as the arithmetic and geometric sequences, it is relatively easy to convert from explicit to recursive definitions and vice versa. In other cases, such as the Fibonacci sequence, this conversion process is much harder.

In this chapter, we will look at a systematic method of finding an explicit formula for a sequence that satisfies a certain type of recursive relationship. In particular, suppose $u_n = a_1 u_{n-1} + a_2 u_{n-2} + \cdots + a_k u_{n-k}$, where $a_1, a_2, \ldots, a_k$ are constants and k is a positive integer. Equations of this form are called *linear homogeneous difference equations of order k with constant coefficients*. The word *linear* refers to the fact that all u_j's appear to the first power. The word *homogeneous* means that there are no terms in the equation other than terms of the sequence. So, for example, $u_n = 2u_{n-1} + 3u_{n-2}$ is a second-order linear homogeneous difference equation with constant coefficients. In contrast, $u_n = 2u_{n-1}^2$ is first-order and homogeneous, but not linear; $u_n = 2u_{n-1} + 5$ is first-order and linear, but not homogeneous due to the presence of the 5.

Any sequence that satisfies the difference equation is said to be a *solution* of the difference equation. Most difference equations have infinitely many solutions. An expression that represents all possible solutions is called the *general solution*. The general solution is usually expressed in terms of one or more arbitrary constants; there is a *particular or specific solution* corresponding to each set of values of these constants.

1. (i) *Verify that* $u_n = 2^n - 1$ *is a solution to* $u_n = 2u_{n-1} + 1$.
    (ii) *Verify that* $u_n = c + d \cdot 2^n$ *is a solution to* $u_n = 3u_{n-1} - 2u_{n-2}$ *for all constants c and d.*

2. *Determine constants c and d such that* $u_n = cn^2 + dn$ *is a solution to* $u_n = u_{n-1} + 2n$.

3. *The only linear homogeneous first-order difference equation is* $u_n = ru_{n-1}$, *where r is a constant. What type of sequence satisfies this equation? Write the general solution to this equation.*

Once we have the general solution to a difference equation (which may or may not be easy to obtain), we can get a particular solution by providing additional information, such as the first term (or several terms) of the sequence. This is not a new idea; we said earlier that to generate a sequence from a recursive definition (which is equivalent to a difference equation), we had to provide one or more initial terms.

In question 1(*i*), $u_n = 2^n - 1$ is a particular solution to $u_n = 2u_{n-1} + 1$ satisfying $u_1 = 1$. It can easily be shown that $u_n = c \cdot 2^n - 1$ is a solution for every value of c. In fact, this is the general solution, although we lack the machinery to prove that. (We'd have to show that there are no other solutions.) In question 1(*ii*), $u_n = c + d \cdot 2^n$ is also the general solution to $u_n = 3u_{n-1} - 2u_{n-2}$.

4. *How much information do we have to specify to derive a particular solution to a first-order difference equation as in question 1(i)?*

5. *How much information do we have to specify to derive a particular solution to a second-order difference equation as in question 1(ii)?*

Since the first-order homogeneous case is not terribly earth-shattering, we turn our attention to the second-order case. The general form for linear homogeneous second-order equations is $u_n = au_{n-1} + bu_{n-2}$, where a and b are constants. In view of our results for first-order equations, we might surmise that a solution to the second-order equation is of the form $u_n = t^n$, for some constant t.

6. *Substitute $u_n = t^n$ in the difference equation and show that t must satisfy the quadratic equation $t^2 = at + b$.*

The quadratic equation above is called the *characteristic equation* for the difference equation. Let's call the roots of the characteristic equation t_1 and t_2. There are three possibilities to consider:

- t_1 and t_2 are real and distinct.
- t_1 and t_2 are real and equal.
- t_1 and t_2 are not real (complex).

For now, we'll consider only the first case. The others are addressed in the Additional Questions in Section 5.4.

Since t_1 and t_2 are the roots of the characteristic equation $t^2 = at + b$, then $u_n = t_1^n$ and $u_n = t_2^n$ are solutions to the difference equation $u_n = au_{n-1} + bu_{n-2}$.

7. *Show that $u_n = c_1 t_1^n + c_2 t_2^n$ is a solution to $u_n = au_{n-1} + bu_{n-2}$ for every c_1 and c_2.*

In fact, if t_1 and t_2 are real and distinct, then $u_n = c_1 t_1^n + c_2 t_2^n$ is the general solution of $u_n = au_{n-1} + bu_{n-2}$, meaning that there are no other solutions. We can get a unique particular solution—that is, solve for c_1 and c_2—by specifying any two terms of the sequence, usually u_1 and u_2 (although in some circumstances, there may be more judicious choices).

8. (i) *Write the general solution to $u_n = 5u_{n-1} - 6u_{n-2}$.*
   (ii) *Write the specific solution to $u_n = 5u_{n-1} - 6u_{n-2}$ satisfying $u_1 = 7$ and $u_2 = 11$.*
   (iii) *Use the difference equation in part (i) to generate the next two terms of the sequence. Show that this agrees with your answer to part (ii).*

9. *Determine the solution to $u_n = 3u_{n-1} + 4u_{n-2}$ satisfying $u_1 = 3$, $u_2 = 17$.*

10. *Can this method be used to solve $u_n = 4u_{n-1} - 4u_{n-2}$? Explain.*

5.1.1. Fibonacci Revisited

Recall the Fibonacci sequence first discussed in Section 1.2. It is defined recursively by $f_1 = 1$, $f_2 = 1$, and $f_n = f_{n-1} + f_{n-2}$ for $n \geq 2$. The first few terms are $\{1, 1, 2, 3, 5, 8, 13, 21, \ldots\}$. When we first encountered this sequence, we said that it is possible to give an explicit formula for f_n in terms of n. We do so now.

Clearly, this is an example of the type of second-order difference equation we've just learned to solve.

11. Show that the roots of the characteristic equation for the Fibonacci sequence are

$$t_1 = \frac{1 + \sqrt{5}}{2} \quad and \quad t_2 = \frac{1 - \sqrt{5}}{2}.$$

Write the general solution to $f_n = f_{n-1} + f_{n-2}$.

12. To simplify the algebra required to solve for the particular solution, we could define f_0 in a way that is consistent with the recursive formula and f_1 and f_2. Do so. Then use f_0 and f_1 to solve for c_1 and c_2.

13. Complete the following theorem, which is known as Binet's formula.

Theorem 5.1 (Binet's Formula)   The nth Fibonacci number is given explicity by $f_n = $ _____ .

The presence of the numbers $t_1 = \dfrac{1 + \sqrt{5}}{2}$ and $t_2 = \dfrac{1 - \sqrt{5}}{2}$ in Binet's formula is somewhat unsettling, since the Fibonacci numbers are all positive integers.

14. Use the Binomial Theorem (Theorem 4.5) to prove that Binet's formula always gives rational values.

Showing that these values are actually integers is more difficult.

Recall that we derived and proved (mostly by induction) a number of interesting relationships about the Fibonacci numbers. Some of these can be easily proven using Theorem 5.1. In the course of proving them, you will need to use the fact that $t_1 - t_2 = \sqrt{5}$, $t_1 + t_2 = 1$, and $t_1 t_2 = -1$.

15. Prove each of the following.

(i) $\displaystyle\sum_{j=1}^{n} f_j = f_{n+2} - 1$ (ii) $f_n^2 = f_{n-1} f_{n+1} + (-1)^{n+1}$

16. Prove that $\displaystyle\lim_{n \to \infty} \frac{f_n}{f_{n-1}} = \frac{1 + \sqrt{5}}{2}.$

The Fibonacci sequence is determined both by the second-order difference equation $f_n = f_{n-1} + f_{n-2}$ and the initial values of f_1 and f_2. Changing any of these may produce an entirely different sequence. Let's see what happens if we change f_2 from 1 to 3. (There is no point setting $f_2 = 2$ since that will just generate the Fibonacci sequence without the first term.) This new sequence is called the *Lucas*[1] sequence. To avoid confusion with the Fibonacci numbers, let g_n equal the nth Lucas number.

17. *Write the first 10 terms of the Lucas sequence.*

18. *Determine an explicit formula for g_n.*

19. *Prove that f_{2n} is always divisible by f_n.*

20. *Prove that* $\lim\limits_{n \to \infty} \dfrac{g_n}{g_{n-1}} = \dfrac{1 + \sqrt{5}}{2}.$

21. *Hypothesize and prove (either directly or by induction) a formula for each of the following.*

$$\text{(i)} \sum_{j=1}^{n} g_j \qquad \text{(ii)} \sum_{j=1}^{n} g_{2j-1}$$

5.2. Linear Function Iteration

Let f be a function and u_0 be a real number. Define a sequence recursively by $u_n = f(u_{n-1})$. Thus, $u_1 = f(u_0)$, $u_2 = f(u_1) = f(f(u_0))$, and so on. In general, $u_n = f^{(n)}(u_0)$, where we use the notation $f^{(k)}(x)$ to mean the outcome obtained by applying the function f k times to the number x. (In other contexts, the same notation is used to refer to the kth derivative of f.) The process of repeatedly applying a function is called *function iteration*.

1. *Enter a positive number in your calculator and repeatedly press the "square root" button. What happens to the results? Try a few other starting numbers. Do you always get the same results?*

2. *Now try the same procedure with the "square" button.*

In some cases, you should have noticed that the sequence of values approaches a fixed value; in other cases, the sequence of values increases without bound.

---

[1] Edouard Lucas (1842–1891) was a French mathematician who did a lot of work in the factorization of large prime numbers. An algorithm known as the Lucas-Lehmer test is still used for checking the primality of Mersenne primes (those of the form $2^p - 1$). It was Lucas who first attached Fibonacci's name to the sequence $\{1, 1, 2, 3, 5, 8, \ldots\}$ after reading about it in Fibonacci's book *Liber Abaci*.

In the first case, we say the sequence *converges*; in the second case, the sequence *diverges*. (This is not the only way in which a sequence diverges; for example, $\{1, -1, 1, -1, \ldots\}$ diverges but remains finite.) The concept of convergence is not new. We saw it in Chapter 1 in our discussion of infinite geometric series and in Chapter 2 in the context of infinite continued fractions.

In this section, we'll restrict our attention to the case in which f is a linear function of the form $f(x) = ax + b$. In other words, we are investigating the sequence defined by $u_n = au_{n-1} + b$. Note that this is a linear first-order difference equation, but it is not homogeneous; thus, the techniques of the last section would need to be modified to solve this equation.

> *3. Suppose $b = 0$. For what values of a would the iteration converge? What would be the limiting value in the cases in which it does converge?*

> *4. Try sufficient experiments to determine the values of a for which the iteration converges when $b \neq 0$. Does the convergence, or lack thereof, depend on b? Does it depend on the initial value u_0?*

> *5. Pick a value of a for which the iteration converges. Determine the limit of the sequence for several values of b. Repeat for several other values of a. Hypothesize an expression for the limit in terms of a and b.*

> *6. What happens when $a = 1$? $a = -1$?*

> *7. Complete the following theorem.*

Theorem 5.2 Let $\{u_n\}$ be a sequence defined recursively by $u_n = au_{n-1} + b$, where a and b are constants.

> (i) If _____, then $\lim_{n\to\infty} u_n$ exists and is equal to _____.
> (ii) If _____, then $\lim_{n\to\infty} u_n$ does not exist.
> (iii) If $a = 1$, then _____.
> (iv) If $a = -1$, then _____.

Theorem 5.2 doesn't quite tell the whole story since $\lim_{n\to\infty} u_n$ may fail to exist for any of several reasons. We'll be more specific a little later. First let's prove what we have so far.

> *8. Express u_1 in terms of a, b, and u_0.*

> *9. Express u_2 in terms of a, b, and u_0.*

> *10. Prove that $u_n = a^n u_0 + \dfrac{b(1 - a^n)}{1 - a}$, if $a \neq 1$. What happens if $a = 1$?*

> *11. Use the result above to determine $\lim_{n\to\infty} u_n$ and the values of a for which it exists. Make sure to consider the cases where $a = \pm 1$.*

Note that the sequence defined explicitly by

$$u_n = a^n u_0 + \frac{b(1 - a^n)}{1 - a}$$

is the specific solution to the nonhomogeneous first-order linear difference equation $u_n = au_{n-1} + b$. The solution consists of two terms. The first term, $a^n u_0$, is the solution to the corresponding homogeneous equation obtained by setting $b = 0$. The second term, $\dfrac{b(1 - a^n)}{1 - a}$, is a particular solution that satisfies the nonhomogeneous equation. Solutions to nonhomogeneous linear difference equations in general are similarly structured. (Solutions to nonhomogeneous systems of linear algebraic equations or nonhomogeneous linear differential equations are also similarly structured.)

Assuming that it exists, we could have obtained $\lim_{n\to\infty} u_n$ directly from the difference equation $u_n = au_{n-1} + b$ itself. Upon taking the limit as n approaches ∞, we get $\lim_{n\to\infty} u_n = a \lim_{n\to\infty} u_{n-1} + b$. Suppose $\lim_{n\to\infty} u_n = U$. Then $\lim_{n\to\infty} u_{n-1} = U$ also; therefore, we have $U = aU + b$.

12. *Solve for U. Does the answer look familiar?*

The equation $U = aU + b$ is a special case of the equation $x = f(x)$ where, in this case, $f(x) = ax + b$. We now have the following definition.

Definition $x*$ is said to be a *fixed point* of the function f if $f(x*) = x*$.

Thus, $x* = \dfrac{b}{1 - a}$ is a fixed point of $f(x) = ax + b$ if $a \neq 1$.

Not every function has a fixed point, and some functions have more than one fixed point. Algebraically, it may be impossible to find the fixed points of a function even if the function has them. For example, there are no algebraic manipulations that can be used to find the fixed point of $f(x) = e^{-x}$—that is, to solve the equation $e^{-x} = x$. However, this function does have a fixed point at approximately $x = 0.56714$.

There are really two questions here: Does a given function f have any fixed points and how do we find them? The first question can be answered graphically.

13. *Describe a graphical method for determining how many fixed points a given function has.*

14. *Use a graph to approximate the values of the fixed points for each of the following.*
    (i) $f(x) = 2 \sin(x)$ (ii) $f(x) = e^x$ (iii) $f(x) = x^3 - x - 3$

15. *Determine (exactly) the fixed points of each of the following.*

   (i) $f(x) = \dfrac{8}{x^2}$ (ii) $f(x) = x^3 - 8x$

16. *Let $f(x) = ax + b$. For which values of a and b does f have*
   (i) *exactly one fixed point?*
   (ii) *no fixed points?*
   (iii) *more than one fixed point?*

As we said earlier, in many cases, it is impossible to find fixed points exactly. So we are forced to approximate them either graphically or numerically. The key to the numerical approximation lies in the following theorem.

Theorem 5.3   Let $\{u_n\}$ be a sequence defined recursively by $u_n = f(u_{n-1})$ for some continuous function f. If $\lim_{n\to\infty} u_n$ exists, then $\lim_{n\to\infty} u_n =$ a fixed point of f.

17. *Prove Theorem 5.3.*

This means that to find the fixed points of f we start with some number u_0 and apply the function f repeatedly. If the process converges—that is, if $\lim_{n\to\infty} u_n$ exists—then it must converge to a fixed point of f. In the case in which f is linear, we know when $\lim_{n\to\infty} u_n$ exists. (Of course, we also know the exact value of the fixed point so we don't have to use the iteration process to find it.) We'll explore the case in which f is nonlinear in the next section.

Notice that Theorem 5.3 does not specify the initial value u_0, so the theorem must be true for every u_0. However, if f has more than one fixed point, then different values of u_0 may lead to convergence to different fixed points. Also, it is clear that if, by chance, we started with $u_0 =$ fixed point, then the iteration will remain at the fixed point.

5.2.1. Cobweb Diagrams

There is a graphical interpretation to the function iteration process. For simplicity, we'll assume f is linear. However, the same argument applies to any function.

The figure below shows the graph of a linear function $f(x) = ax + b$ and the line $y = x$. Let $A(u_0, 0)$ be the point on the x-axis corresponding to the initial term of the iterative sequence. Draw a vertical line from A to the graph of f. Call the intersection point B.

18. *Explain why the coordinates of B are (u_0, u_1).*

Now draw a horizontal line across to the line $y = x$. Call the intersection point C.

19. *Explain why the coordinates of C are (u_1, u_1).*

Continue in this fashion—that is, by drawing vertical lines up to the graph of f and horizontal lines across to the line $y = x$.

20. *What are the coordinates of D and E?*

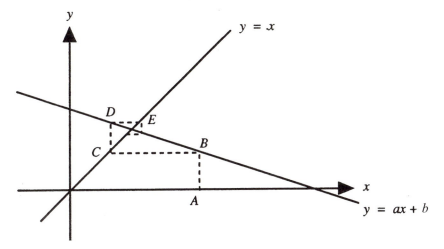

This graphical procedure is called a *cobweb diagram*. As you can see, in this case the process spirals inward toward the fixed point. However, this is not always going to happen.

When drawing cobweb diagrams, it is very important to use the same scale on the x- and y-axes. Otherwise, your results could be misleading.

21. *Draw a cobweb diagram for the function $f(x) = -3x + 4$. Does the process converge?*

Sometimes the cobweb diagram produces not a cobweb, but a "staircase." Nonetheless, the diagram still shows whether the iteration converges.

22. *Draw cobweb diagrams for $f(x) = 3x - 5$ and $f(x) = 0.5x + 4$. In each case, state whether the process converges.*

Definition If the function iteration process converges to a fixed point, we say that the fixed point is *attracting*. If there is a fixed point but the iteration process diverges, we say that the fixed point is *repelling*.

Theorem 5.2 tells us that for linear functions, the iterative process converges if $-1 < a < 1$ and diverges if $a > 1$ or $a < -1$. This should agree with your cobweb diagrams. Now let's consider the case in which $a = 1$.

23. Let f(x) = x + b. How many fixed points does f have? Draw the corresponding cobweb diagrams.

The case in which $a = -1$ is more interesting.

24. Let f(x) = -x + b. Determine the fixed point of f and draw the corresponding cobweb diagram. Compute a few iterates. Is the fixed point attracting or repelling?

25. Show that, for this function, f(f(x)) = x for all b. How does this corroborate your calculations above?

In this last example, the fixed point is said to be *neutral*. The iterative process is said to be *cyclic*.

26. Let f(x) = ax + b, and complete the following summary.

> *(i) If _____ , then f has an attracting fixed point.*
> *(ii) If _____ , then f has a repelling fixed point.*
> *(iii) If _____ , then f has a neutral fixed point.*
> *(iv) If _____ and _____ , then f has no fixed points.*
> *(v) If_____ and _____ , then f has infinitely many fixed*
> *points.*

5.3.   Nonlinear Function Iteration

In the previous section, we showed that every linear function, except those whose slope is 1, has exactly one fixed point. We then showed that, for linear functions whose slope is strictly between -1 and 1, we can determine the fixed point by iterating the function—that is, the fixed point is attracting.

For nonlinear functions, the situation becomes more complicated. Recall that the fixed points of f correspond to the intersections between the graph of f and the line $y = x$. So, nonlinear functions may have more than one fixed point. Some of the fixed points may be attracting, others may be repelling. Recall also that for one specific linear function, the successive iterates oscillated between two numbers. This type of cyclic behavior occurs more frequently with nonlinear functions.

1. Determine a function that has exactly two fixed points.

2. Determine a function other than y = x that has infinitely many fixed points.

3. We claimed in the last section that f(x) = e^{-x} has a fixed point at approximately x = 0.56714. Is this an attracting fixed point? Illustrate with a cobweb diagram.

4. The function $f(x) = 2\cos(x)$ has a fixed point near $x = 1.02$. Is this an attracting fixed point? Illustrate with a cobweb diagram.

5. What must be true for a fixed point of a function to be attracting? [Hint: The answer is a generalization of the linear case.]

A careful statement of the answer to question 5 is contained in the following:

Theorem 5.4   Let $x*$ be a fixed point of a function f. If $|f'(x*)| < 1$, the $x*$ is attracting. If $|f'(x*)| > 1$, then $x*$ is repelling.

Note that if $|f'(x*)| = 1$, there is no conclusion, meaning that $x*$ may be attracting or repelling. Furthermore, the theorem requires us to evaluate the derivative of f at the fixed point. Since we don't know what the fixed point is, it is not possible to do this precisely. However, we can estimate the fixed point graphically and use that estimate in the evaluation of the derivative.

Once we have determined that the fixed point is attracting, we can determine its value more precisely by iterating the function. We should start the iteration with a value that is near the desired fixed point. Indeed, if there are several attracting fixed points, then different starting values will converge to different fixed points.

The proof of Theorem 5.4 makes use of the Mean Value Theorem, which you may have encountered in your calculus course. Essentially, we would need to show that if $|f'(x*)| < 1$, then each iterate is closer to the fixed point than the previous iterate. Moreover, the distances from the fixed point to each iterate decrease by a factor of k, where $k < 1$. Thus, the iteration converges to the fixed point. We won't get involved with the details at this time.

6. Verify that your answers to questions 3 and 4 agree with Theorem 5.4.

This idea of function iteration can be used to find the solutions to equations that otherwise would be difficult to solve. Any equation of the form $g(x) = 0$ can be rewritten in the form $f(x) = x$ for some appropriate function f. Often, there are many ways in which this can be done. All we have to do is choose one so that the function f meets the conditions of Theorem 5.4.

Let $g(x) = x^3 - x - 1$. Upon inspecting the graph, we see that there is one real solution to $g(x) = 0$ between $x = 1$ and $x = 2$. We'll use function iteration to find the root to any desired degree of accuracy. The first step is to rewrite the problem as a fixed point problem—that is, in the form $f(x) = x$.

7. One way to do this is to simply add x to both sides, obtaining $x^3 - 1 = x$. Show numerically that function iteration (beginning with $x = 1$) diverges. Verify this conclusion with Theorem 5.4.

8. *We can rewrite the equation $g(x) = 0$ as a fixed point problem in another way: $\sqrt[3]{x + 1} = x$. Does the function iteration converge? If so, find the desired root to three decimal places. If not, try another form of rewriting the problem.*

9. *Use function iteration to find the smallest positive root of each of the following equations.*
   (i) $\sin(x) = x^2$ (ii) $x = -2 \ln(x)$

5.3.1. Fixed Points and Cycles

As we said earlier, iteration of nonlinear functions can lead to behavior other than simple convergence and divergence. In particular, we can get cyclic behavior similar to what we observed when iterating linear functions with slope -1. In that case, the cycle was of length 2, meaning that successive iterates oscillated between two numbers. It is possible to get cycles of other lengths. It is also possible to get *chaos*, which, loosely speaking, means that the iterates "bounce around" with no apparent pattern, but never "blow up."

It turns out that these different behaviors can be observed by iterating a relatively simple quadratic function $f(x) = bx(1 - x)$, where $0 \leq x \leq 1$ and $b \neq 1$ is a positive constant.

10. *This function has two fixed points. What are they?*

11. *Draw a graph indicating the fixed points. You should consider two cases: $0 < b < 1$ and $b > 1$.*

12. *What happens if $b = 1$?*

From now on, we'll restrict ourselves to the case in which $b > 1$. Our next task is to determine whether the fixed points are attracting or repelling.

13. *Use Theorem 5.4 to show that the nonzero fixed point is attracting for $1 < b < 3$. Draw some cobweb diagrams to support this conclusion.*

The interesting behavior occurs for values of $b > 3$. The following table shows the results of iterating the function for the cases $b = 3.25$, 3.5, and 3.75. (Read the table by rows.)

$b = 3.25$

0.750000	0.609375	0.773621	0.569178	0.796947
0.525924	0.810316	0.499538	0.812499	0.495119
0.812423	0.495274	0.812427	0.495265	0.812427
0.495265	0.812427	0.495265	0.812427	0.495265
0.812427	0.495265	0.812427	0.495265	0.812427
0.495265	0.812427	0.495265	0.812427	0.495265

b = 3.5

0.750000	0.656250	0.789551	0.581561	0.851717
0.442033	0.863239	0.413200	0.848630	0.449599
0.866109	0.405875	0.843991	0.460845	0.869634
0.396797	0.837722	0.475803	0.872951	0.388177
0.831235	0.490992	0.874716	0.383558	0.827544
0.499502	0.874999	0.382815	0.826937	0.500893
0.874997	0.382820	0.826941	0.500884	0.874997
0.382820	0.826941	0.500884	0.874997	0.382820
0.826941	0.500884	0.874997	0.382820	0.826941
0.500884	0.874997	0.382820	0.826941	0.500884

b = 3.75

0.750000	0.703125	0.782776	0.637642	0.866455
0.433914	0.921123	0.272459	0.743344	0.715439
0.763448	0.677231	0.819709	0.554198	0.926485
0.255415	0.713168	0.767097	0.669971	0.829162
0.531198	0.933850	0.231653	0.667461	0.832338
0.523319	0.935461	0.226402	0.656790	0.845313
0.490345	0.937150	0.220873	0.645331	0.858296
0.456089	0.930269	0.243256	0.690310	0.801683
0.596202	0.902795	0.329086	0.827957	0.534165
0.933123	0.234018	0.672200	0.826301	0.538228

14. *Describe the behavior of the iterations when b = 3.25.*

15. *Repeat question 14 for b = 3.5.*

The behavior you observed in these cases is called *cyclic behavior.* More precisely, the numbers x_1 and x_2 form a 2-cycle for the sequence $\{u_n\}$ if, when $u_n = x_1$, then $u_{n+1} = x_2$, and $u_{n+2} = x_1$.

16. *What does this imply about u_{n+3}? u_{n+4}?*

This definition can be easily extended to cycles of arbitrary length.

17. *What is the length of the cycle when b = 3.25? b = 3.5?*

18. *Is there cyclic behavior when b = 3.75?*

Cycles, like fixed points, can be attracting or repelling. If we start near one of the numbers in the cycle and future iterations bring us closer to the numbers in the cycle, then the cycle is attracting (or stable). Otherwise, it is repelling (unstable). The cycles observed for $b = 3.25$ and $b = 3.5$ are both attracting.

In the case of function iteration, a 2-cycle occurs when $u_{n+2} = f(u_{n+1}) = f(f(u_n)) = u_n$. Or, if we let $g(x) = f(f(x))$, then the numbers in the 2-cycle are the fixed points of g that aren't already fixed points of f. If those fixed points are attracting, then the 2-cycle is attracting.

19. *Let $f(x) = bx(1 - x)$. Determine $g(x) = f(f(x))$.*

20. *Use a computer or graphing calculator to draw a graph of g for the cases where $b = 2$, $b = 3$, and $b = 3.5$. How many fixed points does g have in each case?*

Now we'd like to determine a condition under which the cycle is attracting. We know that a function f has an attracting fixed point x^* if $f(x^*) = x^*$ and $|f'(x^*)| < 1$. So for f to have an attracting 2-cycle $\{x_1, x_2\}$, we need to have $|g'(x_1)| < 1$ and $|g'(x_2)| < 1$.

21. *Show that this is equivalent to $|f'(x_1)f'(x_2)| < 1$.*

The next step is to determine the values of a that give rise to an attracting 2-cycle. To do this, we need to solve the equation $g(x) = x$.

22. *Show that this is equivalent to $b^2(1 - x)(1 - bx + bx^2) = 1$, if $x \neq 0$.*

The equation above can be rearranged and factored as

$$(1 - b + bx)(1 + b - bx - b^2x + b^2x^2) = 0.$$

23. *Solve this equation for x.*

There should be three roots to the equation. One of them is the fixed point of f. The other two, if they are real, could be additional fixed points of g.

24. *Determine the values of b for which those additional fixed points are real.*

25. *Use the results of question 21 to show that the 2-cycle is attracting iff $3 < b < 1 + \sqrt{6}$.*

The iterations for $b = 3.5$ seem to indicate the presence of a 4-cycle. Confirming this algebraically is quite a chore since we would have to examine the fixed points of $h(x) = f(f(f(f(x))))$, a 16th-degree polynomial. Nonetheless, it is true that for a small range of b-values, a 4-cycle exists. There are also 8-cycles, 16-cycles, and so on.

For $b = 3.75$, the iterations appear to show no pattern whatsoever. There are no apparent cycles, yet the iterations remain bounded; that is, they don't "blow up." Moreover, had we changed the initial value slightly, we would have gotten an entirely different set of iterates. This is what is known as *chaos*. Chaos theory is an area of mathematical research that has received much attention in recent years.

26. *Chaos occurs for a small range of b-values less than 4. What happens if* $b > 4$?

5.4. Additional Questions

1. Prove that $f_{n-1} + f_{n+1} = g_n$, where f_n is the nth Fibonacci number and g_n is the nth Lucas number.

2. (a) Use Binet's formula to prove that f_{2n} is divisible by f_n for all positive integers n.

(b) Use Binet's formula to show that $\dfrac{f_{3n}}{f_n} = \alpha^{2n} + \alpha^n \beta^n + \beta^{2n}$, where $\alpha = \dfrac{1 + \sqrt{5}}{2}$ and $\beta = \dfrac{1 - \sqrt{5}}{2}$, and argue that $\dfrac{f_{3n}}{f_n}$ is an integer.

3. It is possible to continue the pattern of Problem 2 to prove that f_{kn} is divisible by f_n for all positive integers k and n. Here's another approach.
(a) Use induction to prove that $f_{m+n} = f_{m-1}f_n + f_m f_{n+1}$, for all positive integers m and n. [*Hint:* Show that the statement is true for $n = 1$. Then assume it is true for all $n \leq k$ and show that it is true for $n = k + 1$. Note that this is a somewhat different form of the induction hypothesis.]
(b) Now use induction on k, beginning with $k = 2$, to prove f_{kn} is divisible by f_n for all k.

4. Prove that $\gcd(f_n, f_{n+1}) = 1$ for all n. In other words, prove that successive Fibonacci numbers are relatively prime.

5. Use the fact that $\sum_{j=1}^{n} f_j^2 = f_n f_{n+1}$ to prove that $f_{n+1}^2 = f_n^2 + 3f_{n-1}^2 + 2(f_{n-2}^2 + \cdots + f_1^2)$.

6. We can define f_n for $n \leq 0$ in a manner consistent with pattern for $n > 0$. For example, $f_0 = f_2 - f_1 = 0, f_{-1} = f_1 - f_0 = 1$, and so on. Compute $f_{-2}, f_{-3}, \ldots, f_{-8}$. What is the relationship between f_n and f_{-n}? Prove it.

7. (a) Which Fibonacci numbers are even? Prove your claim.
(b) Which Fibonacci numbers are divisible by 3? Prove your claim.

8. Prove that if f_n is divisible by 2, then $f_{n+1}^2 - f_{n-1}^2$ is divisible by 4.

9. Prove that $\alpha^n = f_n \alpha + f_{n-1}$, where $\alpha = \dfrac{1 + \sqrt{5}}{2}$.

10. Consider the second-order difference equation $u_n = 2u_{n-1} - 5u_{n-2}$, with $u_0 = 2$ and $u_1 = 6$.

   (a) Calculate the next three terms of the sequence.

   (b) Write the corresponding characteristic equation and solve. Notice that the roots are complex conjugates.

   (c) The general solution is $u_n = c_1 r_1^n + c_2 r_2^n$, where r_1 and r_2 are the roots obtained in part (b) Determine c_1 and c_2.

   (d) It is clear from the difference equation that u_n must be an integer for every n. Show that the expression in part (c) supports this.

11. In general, if the roots of the characteristic equation are complex, then the general solution of the difference equation can be expressed as $u_n = cz^n + \overline{cz}^n$, where c and z are complex numbers.

   (a) Suppose $c = r \operatorname{cis}(\varphi)$ and $z = R \operatorname{cis}(\theta)$. Use DeMoivre's Theorem to show that $u_n = 2rR^n \cos(n\theta + \varphi)$.

   (b) If $R > 1$, show that the successive values of u_n are points on a cosine curve with increasing amplitude.

   (c) What happens if $R < 1$?

12. Consider the second-order difference equation $u_n = 4u_{n-1} - 4u_{n-2}$.

   (a) Show that the characteristic equation has two equal real roots.

   (b) In this case, the general solution is not $u_n = c_1 r_1^n + c_2 r_2^n$, where $r_1 = r_2$ are the roots obtained in part (a). Why?

   (c) Show that every sequence of the form $u_n = c_1 r_1^n + c_2 n r_2^n$ is a solution. (In fact, this is the general solution, but proving that takes more ammunition than we have at our disposal.)

13. Suppose we wanted to solve the equation $2 \sin(x) = x$. There is obviously a solution $x = 0$, but there may be others. If there are, no algebraic procedures exist to find them exactly.

   (a) Draw an appropriate graph and estimate the nonzero roots of this equation.

   (b) The roots of this equation can be interpreted as the fixed points of the function $f(x) = 2 \sin(x)$. Consequently, we can use function iteration to find the fixed points, provided the iteration converges. Pick an appropriate initial value and iterate this function. Does the iteration converge? If so, to what number?

   (c) As we proved in Section 5.3, the iteration converges if $|f'(x^*)| < 1$, where x^* is the fixed point. Show that this condition is satisfied for the nonzero fixed points. What about $x = 0$?

14. In question 3 of Section 5.3, we showed that the fixed point of $f(x) = 2 \cos(x)$ is repelling. Determine a value of k such that $f(x) = k \cos(x)$ has an attracting fixed point.

15. We can use function iteration to find cube roots. Consider the equation $x^3 = k$. We can rewrite this as $x = \dfrac{k}{x^2}$; hence, $\sqrt[3]{k}$ is a fixed point of the function $f(x) = \dfrac{k}{x^2}$. So we should be able to find $\sqrt[3]{k}$ by iterating this function, provided the iteration converges.

(a) Show that the iteration does not converge.

(b) Not all is lost. We can rewrite the equation $x^3 = k$ as $x = \sqrt{\dfrac{k}{x}}$, so that $\sqrt[3]{k}$ is a fixed point of the function $f(x) = \sqrt{\dfrac{k}{x}}$. Show that, in this case, the iteration converges.

(c) Use the technique of part (b) to find $\sqrt[3]{15}$ to two decimal places.

(d) Can we use a similar technique to find square roots by writing the equation $x^2 = k$ as $x = \dfrac{k}{x}$?

16. Let $ABCD$ be a rectangle with $AB > AD$. Draw EF parallel to AD so that $AEFD$ is a square. Determine the ratio $AB{:}AD$ if $AB{:}AD = EF{:}EB$.

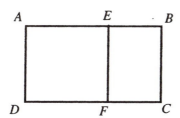

Chapter 6
Additional Topics

There are a number of topics in mathematics that illustrate the basic concepts of this course but do not necessarily fit in any of the previous chapters. We've included a selection of them here. The sections in this chapter are independent of each other and may be read in any order.

6.1. Arithmetic and Geometric Means

We begin with a definition:

Definition Let $x_1, x_2, \ldots, x_n$ be positive numbers. The *arithmetic mean* of these numbers is

$$A(x_1, x_2, \ldots, x_n) = \frac{x_1 + x_2 + \cdots + x_n}{n}.$$

The *geometric mean* of these numbers is

$$G(x_1, x_2, \ldots, x_n) = \sqrt[n]{x_1 x_2 \ldots x_n}.$$

In other words, the arithmetic mean is the ordinary average of the numbers; the geometric mean is the nth root of their product.

1. *Compute each of the following.*
   (i) $A(2, 4, 9)$ (ii) $G(4, 9)$ (iii) $A(A(2, 4), 9)$

2. *Solve each of the following for x.*
   (i) $A(5, 12, x, 2x) = 11$ (ii) $G(x, 2x, 4x) = 10$

3. In the definitions, we required the numbers to be positive. Is this really necessary for the arithmetic mean? for the geometric mean?

4. Prove that $A(x_1 + k, x_2 + k, \ldots, x_n + k) = A(x_1, x_2, \ldots, x_n) + k$ for all k. What is the corresponding result for geometric means?

The arithmetic and geometric means have many interesting properties. We begin by investigating a few simple ones.

5. Suppose $x_1 \leq x_2 \leq \cdots \leq x_n$. Prove that $x_1 \leq A(x_1, x_2, \ldots, x_n) \leq x_n$ and $x_1 \leq G(x_1, x_2, \ldots, x_n) \leq x_n$. In other words, the arithmetic and geometric means of a set of numbers are always greater than or equal to the smallest of the numbers and less than or equal to the largest of the numbers. What happens if all the numbers are equal?

6. Is $A(A(x, y), z) = A(x, y, z)$? If so, prove it. If not, give a counterexample.

7. Is $A(A(x, y), A(z, w)) = A(x, y, z, w)$? If so, prove it. If not, give a counterexample.

8. Repeat questions 6 and 7, replacing "A" with "G."

9. Prove that the natural logarithm of the geometric mean of a set of numbers is equal to the arithmetic mean of the natural logarithms of the numbers.

10. Choose several sets of numbers. In each case, compute the arithmetic and geometric means. Which is bigger?

You should have noticed that the arithmetic mean is at least as big as the geometric mean in every case. In fact, this is true always. A general proof is somewhat difficult, so we will only prove some special cases. This inequality is known as the *arithmetic-geometric-mean (AGM) inequality.*

11. Consider the case in which we have $n = 2$ numbers. Then, we want to show $A(x, y) \geq G(x, y)$ for all x, y. Show that this is equivalent to $(x + y)^2 \geq 4xy$.

12. Show that this, in turn, is equivalent to $(x - y)^2 \geq 0$.

13. Is this last statement true for all x and y? Explain.

Notice the logic we used in questions 11–13. We started with the conclusion of the theorem and worked backward until we got to something that we know to be true. The reason this works is that each step in the backward proof is *equivalent* to the preceding step. That is, the sets of x- and y-values that satisfy each step are exactly the same. In other words, if we denote one statement by P and the next statement by Q, then we can say "P iff Q."

This technique is commonly used. However, to make the proof progress logically, we rewrite the proof so that the conclusion of the theorem is the last statement.

When writing proofs in this manner, we must be very careful that the steps really are equivalent. For example, suppose statement P is "$x = 3$" and statement Q is "$(x + 4)^2 = 49$." Then "if P, then Q" is true, but the converse is false. (Why?) Hence, these steps could not be reversed.

14. Rewrite the proof in questions 11–13 so that the last statement is "Therefore, $A(x, y) \geq G(x, y)$ for all x, y."

15. There actually is a little more to the statement of this theorem. Not only is $A(x, y) \geq G(x, y)$ for all x, y, but $A(x, y) = G(x, y)$ iff $x = y$. Prove this part.

16. The proof of the AGM Inequality for four numbers is not too hard. Try it. You should make use of the result of questions 7 and 8.

The AGM Inequality has many applications. We'll look at a few here.

17. Suppose we want to show that the sum of any positive real number and its reciprocal is always greater than or equal to 2. That is, show that $r + \frac{1}{r} \geq 2$ for every $r > 0$. Write the AGM Inequality with $x = r$ and $y = \frac{1}{r}$. Does this solve the problem? Explain.

18. Suppose we want to find the dimensions of the largest rectangle that can be enclosed with 60 feet of fence. Let x and y be the dimensions of the rectangle, so that the area is xy. Use the AGM Inequality to show that $xy \leq 225$. Hence, the maximum area is no more than 225. To show that the maximum actually is 225, we must find values of x and y that achieve an area of 225. Do so.

Of course, we could have solved these problems with calculus. This is just another approach.

Here's one more example, one for which calculus is not particularly well suited.

19. Prove that $(x + y)(y + z)(x + z) \geq 8xyz$, for all $x, y, z \geq 0$.
[Hint: Write the AGM Inequality separately for each pair of variables; then multiply the three inequalities.]

20. Generalize the result of question 19.

6.1.1. More Means

There are two other means that we can define—the *harmonic mean* and the *root-mean-square*. We will then be able to extend the AGM Inequality to include these means.

21. Suppose you drive from point A to point B at 60 mph and return at 40 mph. What is your average speed for the whole trip?

22. More generally, what would be your average speed if you traveled one direction at x mph and returned at y mph?

Definition   The *harmonic mean* of a set of positive numbers $x_1, x_2, \ldots, x_n$ is defined by

$$H(x_1, x_2, \ldots, x_n) = \frac{n}{\dfrac{1}{x_1} + \dfrac{1}{x_2} + \cdots + \dfrac{1}{x_n}}.$$

23. Show that the answer to question 22 is the harmonic mean of x and y.

24. Suppose $x_1 \le x_2 \le \cdots \le x_n$. Prove that $x_1 \le H(x_1, x_2, \ldots, x_n) \le x_n$.

Definition   The *root-mean-square* of a set of positive numbers $x_1, x_2, \ldots, x_n$ is defined by

$$R(x_1, x_2, \ldots, x_n) = \sqrt{\frac{x_1^2 + x_2^2 + \cdots + x_n^2}{n}}.$$

Note that the term *root-mean-square* is quite descriptive since it is the square root of the average (or mean) of the squares of the numbers.

25. Suppose $x_1 \le x_2 \le \cdots \le x_n$. Prove that $x_1 \le R(x_1, x_2, \ldots, x_n) \le x_n$.

26. For the sets of numbers you used in question 10, compute the harmonic mean and root-mean-square. Do all four means always come out in the same order?

27. Prove your observation for n = 2 numbers.

28. Show that all $A(x, y) = G(x, y) = H(x, y) = R(x, y)$ iff $x = y$.

The harmonic mean has applications in physics. For example, if two resistors are connected in parallel, the total resistance of the circuit is

$$\frac{1}{2} H(r_1, r_2) = \frac{1}{\dfrac{1}{r_1} + \dfrac{1}{r_2}}.$$

The root-mean-square has applications in statistics. The sample standard deviation s of a set of numbers is

$$s = \sqrt{\frac{\sum_{j=1}^{n}(x_j - \bar{x})^2}{n - 1}},$$

where $\bar{x}$ is the arithmetic mean (average) of the numbers. Thus,

$$s = \sqrt{\frac{n}{n - 1}} R(x_1 - \bar{x}, x_2 - \bar{x}, \ldots, x_n - \bar{x}).$$

It is a measure of how spread out the numbers are.

6.2. Greatest Integer Function

We begin with a definition.

Definition The *greatest integer function* $f(x) = [x]$ is the largest integer that is less than or equal to x.

In other words, $[x]$ is the unique integer satisfying $x - 1 < [x] \leq x$.

1. *Determine each of the following.*
   (i) $[2.3]$ (ii) $[5]$ (iii) $[-3.4]$ (iv) $[-\sqrt{2}]$

2. *Determine all the values of x that make each of the following true.*
   (i) $[x] = 4$ (ii) $[x] = -4$ (iii) $[2x] = 6$
   (iv) $3[x] = 7$ (v) $[x^2] = 2$ (vi) $[-x] = 1$

3. *Draw a graph of $y = [x]$ for $-2 \leq x \leq 3$.*

4. *Is $[x] + [x] = [2x]$ for all x? If not, for which x-values is it true?*

5. *Is $[x] + [-x] = 0$ for all x? If not, for which x-values is it true?*

Here are some elementary facts about the greatest integer function, which we state without proof.

Theorem 6.1 Let x and y be real numbers and let m be an integer.

  (i) If $m \leq x$, then $m \leq [x]$.
 (ii) If $m > x$, then $m \geq [x] + 1$.
(iii) If $x \leq y$, then $[x] \leq [y]$.
(iv) $[x + m] = [x] + m$.

(v) If $x = m + z$, where $0 \leq z < 1$, then $m = [x]$.

Recall that the Division Algorithm for integers (Theorem 1.2) says that given any two positive integers a and b, there exists unique nonnegative integers q and r such that $a = qb + r$, where $0 \leq r < b$.

6. *Prove that* $q = \left[\dfrac{a}{b}\right]$.

It is convenient to define the *fractional part* of a real number by $\{x\} = x - [x]$. Clearly, $0 \leq \{x\} < 1$ for all x.

7. *Determine each of the following.*
   (i) $\{3.6\}$ *(ii)* $\{-3.6\}$

8. *Is $[x + y] = [x] + [y]$? If so, prove it; otherwise, fix it so that it is true.*

9. *Is $\{x + y\} = \{x\} + \{y\}$? If so, prove it; otherwise, fix it so that it is true.*

10. *Create sufficient examples to convince yourself that*

$$\left[\frac{[x]}{n}\right] = \left[\frac{x}{n}\right]$$

for all real numbers x and positive integers n.

11. *To prove the result in question 10, we begin by observing that*

$$\left[\frac{x}{n}\right] \leq \frac{x}{n} < \left[\frac{x}{n}\right] + 1.$$

Why is this true?

12. *Multiply the inequality above by n. Then argue that*

$$n\left[\frac{x}{n}\right] \leq [x] < n\left[\frac{x}{n}\right] + n.$$

Use Theorem 6.1.

13. *Now divide by n and complete the proof. (Remember that $\left[\dfrac{x}{n}\right]$ is an integer.)*

Proofs involving the greatest integer function sometimes involve looking at disjoint cases. For example, suppose we wanted to prove that $[x] + [x + \frac{1}{2}] = [2x]$ for all x. Let $x = [x] + \{x\}$. There are two cases to consider.

14. The first case is when $\{x\} < 0.5$. If $\{x\} < 0.5$, then $[x] = [x + \frac{1}{2}]$. Complete this case.

15. The other case occurs when $\{x\} \geq 0.5$. Complete this case.

16. Prove that $[x] + [x + \frac{1}{3}] + [x + \frac{2}{3}] = [3x]$.

We'll close this section by looking at an interesting application of the greatest integer function. Let n be a positive integer and p be a prime. Determine the exponent of p in the canonical representation of $n!$.

Certainly there is a factor of p for each multiple of p that is less than or equal to n.

17. How many such multiples are there?

18. This does not represent the entire answer since some of the integers less than or equal to n are divisible by p^2 and, thus, we pick up additional factors of p from those integers. How many of them are there?

19. Continuing in this fashion, we also have to account for those integers divisible by p^3, p^4, and so on. How many of each of those are there?

20. Conclude that the exponent of p in the canonical representation of n! is

$$\sum_{k=1}^{\infty} \left[\frac{n}{p^k} \right].$$

How do we know this sum is finite?

21. Test the formula by finding the canonical representations of 6!, 8!, and 13! and determining the exponent of each prime factor.

One interesting thing we can now prove is that the binomial coefficients $\binom{n}{r}$ are always integers. (Of course, the combinatorial interpretation as the number of subsets of size r from a set with n elements ensures that they must be integers. What we will show is that the computational formula $\frac{n!}{r!(n-r)!}$ given by Theorem 4.4 always produces an integer.)

To prove this, we'll show that the exponent of each prime in the canonical representation of the numerator is at least as big as the exponent of that prime in the denominator. We'll need to use the fact that $[x + y] \geq [x] + [y]$ for all x and y.

22. Let p be a prime in the canonical representation of n! and let k be a positive integer. Show that

$$\left[\frac{n}{p^k}\right] \geq \left[\frac{r}{p^k}\right] + \left[\frac{n-r}{p^k}\right].$$

23. Upon adding the inequalities above for each k = 1, 2, 3, . . . , we get

$$\sum_{k=1}^{\infty}\left[\frac{n}{p^k}\right] \geq \sum_{k=1}^{\infty}\left[\frac{r}{p^k}\right] + \sum_{k=1}^{\infty}\left[\frac{n-r}{p^k}\right].$$

How does this prove the result?

24. Use this result to prove that the product of r consecutive integers is always divisible by r!.

6.3.   Sums and Differences

Let $\{u_n\}$ be a sequence of real numbers. Define a new sequence $\{\Delta u_n\}$ by $\Delta u_n = u_{n+1} - u_n$, $n = 1, 2, 3, \ldots$. This new sequence is called the *difference* of the original sequence.

1. Compute the first five terms of the difference of {1, 4, 8, 13, 10, 6, 12, . . .}.

2. Derive a formula for Δu_n for each of the following.
  (i) $u_n = n^2$ (ii) $u_n = 3n + 7$ (iii) $u_n = 2^n$
  (iv) $u_n = \dfrac{1}{n}$ (v) $u_n = n!$

Verify that your formulas are correct by computing several terms of u_n and Δu_n.

We can draw a graph of the sequence by plotting points whose coordinates are (n, u_n). Since the sequence is defined only for integer values of n, we do not "connect the dots."

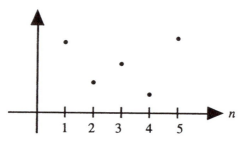

3. What is the geometric interpretation of Δu_n in the context of this graph? Is this reminiscent of something you learned in calculus?

We can define higher-order differences as well. For instance, the second difference $\Delta^2 u_n$ is $\Delta(\Delta u_n)$.

4. *Express $\Delta^2 u_n$ in terms of the original sequence.*

5. *Compute $\Delta^2 u_n$ for each of the sequences in question 2.*

6. *Repeat question 4 for $\Delta^3 u_n$ and $\Delta^4 u_n$. Do you see a general pattern emerging?*

7. *What is the graphical interpretation of $\Delta^2 u_n$?*

Now we'll prove some simple theorems about the difference operator. Many of these theorems are similar (but not necessarily identical) to the corresponding theorems about derivatives.

8. *Prove that $\Delta(u_n + v_n) = \Delta u_n + \Delta v_n$.*

9. *Prove that $\Delta(k u_n) = k \Delta u_n$, where k is a constant.*

10. *Construct a counterexample to show that $\Delta(u_n v_n) \neq \Delta u_n \Delta v_n$.*

11. *By analogy to the product rule for derivatives, we might think that $\Delta(u_n v_n) = v_n \Delta u_n + u_n \Delta v_n$. Is this correct? If so, prove it. If not, fix it.*

12. *Derive and prove a formula for*

$$\Delta\left(\frac{u_n}{v_n}\right).$$

13. *Suppose $u_n = n^r$, where r is a positive integer. Derive a formula for Δu_n in the case where $r = 1$; $r = 2$; $r = 3$; $r = 4$. Generalize.*

14. *Use the theorems above to compute Δu_n if $u_n = 3n^2 + 4n + 8$.*

6.3.1. Summations and Antidifferences

15. *Suppose you are given the difference of a sequence. Can you reconstruct the original sequence? If not, what other information do you need?*

16. *Show that $u_n = u_1 + \sum_{k=1}^{n-1} \Delta u_k$, for $n \geq 2$.*

It makes sense to define an *antidifference* of a sequence in much the same way we define the antiderivative of a function; that is, v_n is an antidifference of u_n if $u_n = \Delta v_n$ for all n. We'll write $v_n = \Delta^{-1} u_n$.

17. *By saying "an" antidifference, rather than "the" antidifference, we imply that there may be more than one for any given sequence. How many are there and how are they related?*

18. *Find the first few terms of all the antidifferences of the sequence*

$$\{3,\ -2,\ 5,\ 7,\ -6,\ 4,\ \ldots\}.$$

19. *Finding antidifferences symbolically is not easy, especially in view of the not-so-simple results in question 13. However, in some cases, we can try to guess or "wave our hands" to come up with formulas. Find $\Delta^{-1} u_n$ for each of the following.*
    (i) $u_n = n$ (ii) $u_n = 2^n$

20. *Show that if $u_n = b^n$, then*

$$\Delta^{-1} u_n = \frac{b^n}{b-1}.$$

21. *The equation in question 16 can be rewritten as $\Delta^{-1} u_n - \Delta^{-1} u_1 = \sum_{k=1}^{n-1} u_k$. Interpret the summation on the right as an appropriate area on the plot of the sequence. Does this equation remind you of something in calculus?*

The equation in question 21 implies that we can evaluate some summations by finding an antidifference. First let's adjust the limits of the summation so that we find the sum of the first n terms (not $n - 1$) of the sequence. We do this by replacing the $n - 1$ in the summation by n and the n in the antidifference by $n + 1$. That is,

$$\Delta^{-1} u_{n+1} - \Delta^{-1} u_1 = \sum_{k=1}^{n} u_k.$$

For example, suppose we want to evaluate $\sum_{k=1}^{n} 2^k$. In question 19(ii), we showed that $\Delta^{-1}(2^k) = 2^k$. Therefore, $\sum_{k=1}^{n} 2^k = 2^{n+1} - 2^1 = 2^{n+1} - 2$. Since the summation is just a geometric series, we can easily verify that the result is correct using Theorem 1.4, or we can prove it by induction. (Note that the first term is missing since the summation starts with $k = 1$, not 0.)

Similarly, we have shown that $\Delta^{-1}(2k + 1) = k^2$. Hence, $\sum_{k=1}^{n}(2k + 1) = (n + 1)^2 - 1^2 = n^2 + 2n$. Here, the summation is an arithmetic series, so we can use Theorem 1.3, or induction, to verify the result.

22. *Use this procedure and the result of question 2(v) to evaluate $\sum_{k=1}^{n} k(k!)$.*

As with antiderivatives, our ability to find antidifferences symbolically is extremely limited. (In fact, it's much more limited for antidifferences.) There are a few techniques that can help us in this regard.

In question 11, we proved that $\Delta(u_n v_n) = v_{n+1}\Delta u_n + u_n \Delta v_n$.

23. *Rearrange terms and then take the antidifference of both sides to show that $\Delta^{-1}(u_n \Delta v_n) = u_n v_n - \Delta^{-1}(n_{n+1} \Delta u_n)$. Does this look familiar?*

We'll call this technique *summation by parts*. (What else?)

Consider the summation $\sum_{k=1}^{n} k \cdot 2^k$. To evaluate this sum, let $w_n = n \cdot 2^n$. According to the formula following question 21, $\sum_{k=1}^{n} k \cdot 2^k = \Delta^{-1} w_{n+1} - \Delta^{-1} w_1$. We'll use summation by parts to find $\Delta^{-1}(n \cdot 2^n)$. We already know $\Delta^{-1}(2^n) = 2^n$.

24. *Use summation by parts with $u_n = n$ to show that*

$$\Delta^{-1}(n \cdot 2^n) = n \cdot 2^n - \Delta^{-1}(2^{n+1}) = n \cdot 2^n - 2^{n+1}.$$

It follows that

$$\sum_{k=1}^{n} k \cdot 2^k = \Delta^{-1} w_{n+1} - \Delta^{-1} w_1 = (n+1)2^n - 2^{n+2} - (-2) = 2^{n+1}(n-1) + 2.$$

25. *Check this formula for $n = 3$, 4 and 5.*

26. *Prove the formula by induction.*

6.4. Diophantine Equations and the Chinese Remainder Theorem

Consider the equation $3x + y = 7$. If x and y are real numbers, then there are infinitely many solutions to this equation.

1. *Draw a graph showing all solutions to this equation.*

Now suppose we restrict x and y to integer values. This eliminates many of the solutions we found previously.

2. *Determine four integer solutions to the equation above. Locate those solutions on your graph.*

In general, an equation whose variables are restricted to integer values is called a *Diophantine equation*. In this section, we'll consider linear Diophantine equations in two variables—that is, those of the form $ax + by = c$ where a, b, and c are integers.

It is not clear that an equation of this form has any solutions at all. For example, consider the equation $3x + 6y = 10$.

3. *Show that this equation has no solutions.*

4. *For which values of c does $3x + 6y = c$ have a solution?*

5. *For which values of c does $10x + 15y = c$ have a solution?*

We're now ready to state a necessary condition for a linear Diophantine equation to have a solution.

6. Complete the following theorem.

Theorem 6.2   If $ax + by = c$ has a solution, then _____ .

To prove this theorem, observe that $a = dm$ and $b = dn$, where $d = \gcd(a, b)$ and m and n are relatively prime.

7. Complete the proof of Theorem 6.2.

Theorem 6.2 essentially tells us when the equation $ax + by = c$ does *not* have a solution. We would really like to have a condition that guarantees that the equation *does* have a solution. In other words, we would like to claim that the converse of Theorem 6.2 is true also.

The converse is indeed true. To prove it, we first need the following lemma, which we state without proof.

Lemma 6.1   If a and b are integers, not both zero, and $d = \gcd(a, b)$, then there exist integers u and v such that $d = au + bv$.

8. Try a few examples so that you understand Lemma 6.1.

9. Complete and prove Theorem 6.3, which is the converse of Theorem 6.2.

Theorem 6.3   If _____ , then there exist integers x and y such that $ax + by = c$.

Next, we'd like to show that if the Diophantine equation $ax + by = c$ has a solution, then it has infinitely many solutions. Consider the equation $4x + 10y = 22$. One solution is $x = 3$, $y = 1$.

10. What is the next larger value of x for which there is a solution? What is the corresponding y-value? What is the solution with the next larger x-value?

11. Show that $x = 3 + 5t$, $y = 1 - 2t$ is a solution for every integer value of t. How do the coefficients 5 and −2 relate to the coefficients of the Diophantine equation?

12. In general, suppose (x_0, y_0) is a solution of the equation $ax + by = c$. Describe a method of generating infinitely many solutions. Does your method generate all solutions?

13. Complete the following corollary.

Corollary 6.1   If a and b are relatively prime, then $ax + by = c$ has a solution, for _____ .

6.4.1. Linear Congruences

Linear Diophantine equations of the form $ax + by = c$ can be reformulated in terms of congruences and modular arithmetic.

14. Prove that $ax - my = c$ iff $ax \equiv c \bmod m$.

We'll now concentrate on determining under what conditions the linear congruence $ax \equiv c \bmod m$ has a solution. More specifically, given a and c, is there a value of x such that $ax = c$, where the arithmetic is understood to be done in $Z_m = \{0, 1, \ldots, m - 1\}$? Let's begin by creating multiplication tables in Z_5 and Z_6. (You might wish to refer to Section 1.1 for a review of congruences and modular arithmetic.)

15. Complete the following tables.

Z_5

•	0	1	2	3	4
0	0	0	0	0	0
1	0	1	2	3	4
2	0	2	4	1	3
3					
4					

Z_6

•	0	1	2	3	4	5
0	0	0	0	0	0	0
1	0				4	
2	0	2				
3	0			3		
4	0					2
5	0		4			

16. Use the tables above to determine all solutions to each of the following.
 (i) $2x \equiv 3 \bmod 5$ *(ii)* $2x \equiv 5 \bmod 6$
 (iii) $5x \equiv 4 \bmod 6$ *(iv)* $2x \equiv 4 \bmod 6$
 (v) $3x \equiv 3 \bmod 6$

17. *Determine all solutions to each of the following.*
    (i) $3x \equiv 4 \bmod 7$ *(ii)* $2x \equiv 4 \bmod 8$

You should have noticed that in some cases there is no solution. In other cases, there is a unique solution in Z_m, and in the remaining cases, there is more than one solution. Our goal is to determine the number of solutions.

In general, the number of solutions of $ax \equiv c \bmod m$ is the number of times "c" appears in the "a" row (or column) in the multiplication table for Z_m. In the table for Z_5, every element of Z_5 appears exactly once in each nonzero row.

18. *What does this mean about the solutions to $ax \equiv c \bmod 5$?*

The situation in Z_6 is quite different. In some rows, every element appears exactly once; in others, some elements appear more than once and others do not appear at all.

19. *In which rows of the Z_6 multiplication table does every element appear exactly once? What is special about those rows? Does this agree with the corollary to Theorem 6.3? Explain.*

20. *Show that the remaining rows agree with Theorem 6.3 itself.*

Notice that those elements that do appear more than once in any given row all appear the same number of times in that row. For instance, in row 2, the elements 0, 2, and 4 all appear twice; in row 3, the elements 0 and 3 each appear three times.

21. *For the values of c for which there are multiple solutions, how is the number of distinct solutions to $ax \equiv c \bmod 6$ related to a?*

We're now ready to state the main result.

22. *Complete the following theorem.*

Theorem 6.4   The linear congruence $ax \equiv c \bmod m$ has a solution iff _____ , where $d = \gcd(a, m)$. Moreover, if there is a solution, then there are _____ distinct solutions modulo m.

The first sentence in this theorem is merely a restatement of Theorems 6.2 and 6.3 and, as such, needs no further justification. The second sentence requires more work.

We've previously shown that if (x_0, y_0) is a solution to the Diophantine equation $ax - my = c$, then every other solution is of the form $x = x_0 + \frac{m}{d}t, y = y_0 +$

$\frac{a}{d}t$, where $d = \gcd(a, m)$ and t is an integer. Consider the d solutions whose x-values correspond to $t = 0, 1, 2, \ldots, d - 1$; that is,

$$\left\{ x_0, x_0 + \frac{m}{d}, x_0 + \frac{2m}{d}, \ldots, x_0 + \frac{(d - 1)m}{d} \right\}.$$

We need to show two things: that no two of these are congruent mod m and that every other solution is congruent to one of these mod m.

To prove the first of these, we need the following lemma, which we state without proof.

Lemma 6.2 If $kr \equiv ks \bmod n$, then $r \equiv s \bmod(\frac{n}{g})$, where $g = \gcd(k, n)$.

Now pick two arbitrary members of the set of x-values, say $x_0 + \frac{m}{d}t_1$ and $x_0 + \frac{m}{d}t_2$, where $0 \le t_1 < t_2 \le d - 1$. Assume that they are congruent mod m.

23. *Use Lemma 6.2 to show that* $t_1 \equiv t_2 \bmod d$. *This is impossible. Why?*

Thus, no two of the x-values are congruent mod m. Now consider any other solution of the form $x_0 + \frac{m}{d}t$ for some $t \ge d$. By the Division Algorithm, $t = qd + r$, where $0 \le r \le d - 1$.

24. *Show that* $x_0 + \frac{m}{d}t \equiv (x_0 + \frac{m}{d}r) \bmod m$. *Why does this complete the proof?*

25. *Solve each of the following.*
    (i) $4x \equiv 8 \bmod 12$ (ii) $8x \equiv 4 \bmod 18$

6.4.2. Chinese Remainder Theorem

The solutions to the linear congruence $x \equiv c \bmod m$ give us all the numbers that leave a remainder of c upon division by m. This, by itself, is not a particularly interesting problem. However, suppose we want to find all numbers that satisfy more than one congruence simultaneously.

For example, let's find all numbers that leave a remainder of 2 upon division by 6 and a remainder of 3 upon division by 5. This is equivalent to solving the congruences $x \equiv 2 \bmod 6$ and $x \equiv 3 \bmod 5$ simultaneously.

From the first congruence, $x = 6m + 2$, for some integer m.

26. *Substitute this expression for x into the second congruence. Does the resulting equation have a solution modulo 5? Is it unique?*

27. *Write an expression that gives all the values of m that satisfy the equation in question 26.*

28. Substitute this in $x = 6m + 2$ and show that all the values of x that satisfy the two congruences simultaneously are of the form $x \equiv 8 \bmod 30$.

29. Determine all solutions to $x \equiv 5 \bmod 6$ and $x \equiv 3 \bmod 7$.

From these examples, we might be led to believe that the congruences $x \equiv c_1 \bmod m_1$ and $x \equiv c_2 \bmod m_2$ have a unique solution of the form $x \equiv c \bmod m_1 m_2$ for some c. In fact, this is not quite true.

30. Show that the solutions to $x \equiv 3 \bmod 4$ and $x \equiv 1 \bmod 6$ are of the form $x \equiv 7 \bmod 12$.

The correct result, which is known as the Chinese Remainder Theorem,[1] can be stated as follows:

Theorem 6.5 (Chinese Remainder Theorem)   If m_1 and m_2 are relatively prime, then the congruences $x \equiv c_1 \bmod m_1$ and $x \equiv c_2 \bmod m_2$ have a unique solution of the form $x \equiv c \bmod m_1 m_2$.

31. Prove the Chinese Remainder Theorem for the case of two congruences.

Although we have stated the Chinese Remainder Theorem for two congruences, it can be extended to more than two, provided that every pair of moduli is relatively prime.

6.5. Pythagorean Triplets

We are all familiar with the Pythagorean Theorem, which states that if x, y, and z are the sides of a right triangle, then $x^2 + y^2 = z^2$.

Any triplet of positive integers satisfying the Pythagorean Theorem is called a *Pythagorean triplet*. The most common (and smallest) Pythagorean triplet is $(3, 4, 5)$; other well-known ones are $(5, 12, 13)$ and $(8, 15, 17)$.

1. Prove that if (x, y, z) is a Pythagorean triplet, then for any positive integer k, (kx, ky, kz) is also a Pythagorean triplet.

In view of the result in question 1, we'll restrict our attention to *primitive* Pythagorean triplets—that is, those in which x, y, and z are relatively prime or, equivalently, $\gcd(x, y, z) = 1$. Thus, $(3, 4, 5)$ is a primitive Pythagorean triplet whereas $(6, 8, 10)$ is not.

---

[1]The name arises from the fact that in the first century A.D. the Chinese mathematician Sun-Tsu posed the following problem: Find a number that leaves remainders of 2, 3, and 2 when divided by 3, 5, and 7, respectively. As we said, this is equivalent to a problem in simultaneous congruences.

Our goal is to find all (primitive) Pythagorean triplets. In other words, we want to find all relatively prime, positive integer solutions to the quadratic Diophantine (Pythagorean) equation $x^2 + y^2 = z^2$.

Although our definition of "primitive" just requires that $\gcd(x, y, z) = 1$, we can also prove that x, y, and z are *pairwise* relatively prime. That is, we can show $\gcd(x, y) = \gcd(y, z) = \gcd(x, z) = 1$. To see why, suppose to the contrary that $\gcd(x, y) = d > 1$. Then there must be some prime p such that x and y are divisible by p.

2. *Show that z^2 and, consequently, z are divisible by p. Where is the contradiction?*

3. *Argue that x and y cannot both be even.*

The same argument applies to the other two pairs.

It is also true that x and y cannot both be odd. To prove this, suppose that $x = 2m + 1$ and $y = 2n + 1$.

4. *Show that $z^2 \equiv 2 \bmod 4$. Why is this impossible?*

It follows that either x is odd and y is even or vice versa. By symmetry, it doesn't really matter which of them is odd, so for the remainder of this discussion we'll assume that x is odd. Of course, this also implies that z is odd.

The Pythagorean triplets (3, 4, 5), (5, 12, 13), (7, 24, 25), (9, 40, 41), ... all follow a certain pattern.

5. *State the pattern in general and prove that it works. Generate the next two triplets in the pattern.*

6. *Show that there are Pythagorean triplets that don't conform to this pattern.*

The complete solution to the Pythagorean equation $x^2 + y^2 = z^2$ is given by the following theorem.

Theorem 6.6   All positive solutions of the equation $x^2 + y^2 = z^2$ with x odd and $\gcd(x, y, z) = 1$ are given by $x = s^2 - t^2$, $y = 2st$, and $z = s^2 + t^2$, where s and t are relatively prime integers such that $s > t$ and s and t are not both odd or both even.

Before we prove this theorem, let's do some calculations.

7. *Find the values of s and t that correspond to the Pythagorean triplets (3, 4, 5), (5, 12, 13), (8, 15, 17), and (20, 21, 29). Verify that those values of s and t satisfy the conditions of the theorem.*

8. *Pick some other pairs of values of s and t and compute the corresponding Pythagorean triplets.*

The proof consists of several parts. First we have to show that all x, y, and z of the given form satisfy the Pythagorean equation and that they are relatively prime. Then we have to show that there are no other solutions of a different form.

9. Show that $x = s^2 - t^2$, $y = 2st$, and $z = s^2 + t^2$ satisfy the Pythagorean equation. Also show that x and z are odd and y is even.

10. Suppose that $\gcd(x, z) = d > 1$. Then there exists a prime p such that x and z are both divisible by p. Show that $2s^2$ and $2t^2$ must be divisible by p.

11. Argue that p must be odd and, consequently, s^2 and t^2 are divisible by p. Why is this a contradiction?

It follows from questions 10 and 11 that $\gcd(x, z) = 1$. A similar argument can be used to show that the other pairs are relatively prime. Hence, the Pythagorean triplet prescribed by Theorem 6.6 is primitive.

Now let's show that there are no solutions to the Pythagorean equation other than those give in Theorem 6.6. Let (x, y, z) be a primitive Pythagorean triplet. Since x and z are both odd, $z - x$ and $z + x$ are both even. Let $z - x = 2u$ and $z + x = 2v$, where u and v are integers.

12. Express z and x in terms of u and v.

13. Argue by contradiction that u and v must be relatively prime.

14. Argue that one of u and v is even and the other is odd.

15. Show that $uv = \left(\dfrac{y}{2}\right)^2$.

The next step hinges on the following lemma whose proof is straightforward.

Lemma 6.3 If m and n are relatively prime integers whose product is a perfect square, then m and n must each be perfect squares.

16. Is Lemma 6.3 true if m and n are not relatively prime? Give an example.

17. It follows from Lemma 6.3 that u and v must be perfect squares. Complete the proof of Theorem 6.6.

We now know how to generate all primitive Pythagorean triplets.[2]

---

[2]Pythagoras, who lived in the sixth century B.C., is generally credited with discovering and proving the theorem named for him. However, there is strong evidence that the Babylonians knew how to calculate Pythagorean triplets more than 1000 years before Pythagoras' birth. They apparently also knew the results stated in this section. For more details, see *An Introduction to the History of Mathematics*, by Howard Eves (Holt, Rinehart and Winston, Inc., 1969).

6.5.1. Fermat's Last Theorem

Solutions to the Pythagorean equation $x^2 + y^2 = z^2$ have been known for at least 2500 years. In the seventeenth century, Pierre de Fermat considered the more general equation $x^n + y^n = z^n$, where n is a positive integer. Fermat claimed that if $n \geq 3$, then this equation has no positive integer solutions. He also claimed to be able to prove it, but the "margin (of his notebook) was too small to contain it."

It turns out that it is extraordinarily difficult to prove this claim, which has come to be known as Fermat's Last Theorem. Mathematicians worked on it unsuccessfully for 350 years, although a lot of progress had been made.[3] Finally, in 1993, Andrew Wiles announced that he had a complete proof. Although his proof had a few holes in it, those holes were eventually patched and it is now generally accepted that Fermat's Last Theorem has been proved. Given that the proof uses mathematics that hadn't even been dreamed of in Fermat's time, it is hard to believe that Fermat had a proof. However, it is possible that there is a much simpler proof lurking somewhere that still has not been found, one that Fermat may have indeed discovered.

Needless to say, we are not going to prove Fermat's Last Theorem here. We will, however, prove a special case where $n = 4$. That is, we will show that the equation $x^4 + y^4 = z^4$ has no positive integer solutions.

18. Why do we require the solutions to be <u>positive</u> integers?

We begin by considering a slightly different equation—$x^4 + y^4 = z^2$ or, equivalently, $(x^2)^2 + (y^2)^2 = z^2$. This is a variation of the Pythagorean equation and, consequently, by Theorem 6.6 every solution must be of the form $x^2 = s^2 - t^2$, $y^2 = 2st$, $z = s^2 + t^2$ for some relatively prime positive integers s and t such that s and t are not both odd or both even. Note that this implies that x^2 and z are odd and y^2 is even.

Suppose s is even and t is odd.

19. Show that $s^2 - t^2 \equiv 3 \bmod 4$ but $x^2 \equiv 1 \bmod 4$.

---

[3]Progress in the form of proving special cases was made beginning with Euler, who proved the theorem for $n = 3$. The Frenchmen Peter Gustav Dirichlet and Adrien-Marie Legendre proved the case $n = 5$ in 1825; Gabriel Lamé proved the case $n = 7$ in 1839. German mathematician Ernst Kummer claimed to have a complete proof in 1843. Dirichlet found an error in the proof, which resulted from Kummer having made assumptions about unique factorization that turned out to be false. Later, Kummer successfully proved Fermat's Last Theorem for a large class of prime values of n. In the process, he discovered many new algebraic ideas.

Eventually, what was needed was an entirely new approach. This occurred in the 1980s when it was shown that Fermat's Last Theorem was related to the study of elliptic curves, that is, curves defined by equations of the form $y^2 = x^3 + ax + b$, where a and b are integers.

Since we have a contradiction, it must be that s is odd and t is even. Then, upon letting $t = 2r$, we have $y^2 = 4sr$ or $\left(\dfrac{y}{2}\right)^2 = sr$.

20. *Argue that s and r are relatively prime and that s and r are both perfect squares.*

Since $x^2 = s^2 - t^2$, then (x, t, s) is a primitive Pythagorean triplet satisfying $x^2 + t^2 = s^2$. Thus, from Theorem 6.6, there exist relatively prime positive integers u and v such that $x = u^2 - v^2$, $t = 2uv$, and $s = u^2 + v^2$. The same argument used in question 20 shows that u and v must be perfect squares.

Now define integers x_1, y_1, and z_1 such that $u = x_1^2$, $v = y_1^2$, and $s = z_1^2$.

21. *Show that $x_1^4 + y_1^4 = z_1^2$.*

22. *Convince yourself that $0 < z_1 \le z_1^2 = s \le s^2 < s^2 + t^2 = z$.*

Thus, starting with some solution (x, y, z) satisfying $x^4 + y^4 = z^2$, we have produced another solution (x_1, y_1, z_1) with $0 < z_1 < z$. We could repeat this process, producing another solution with a *strictly smaller* z-value.

23. *Why can't this process go on forever?*

24. *Explain how we have now shown that $x^4 + y^4 = z^2$ has no positive integral solutions.*

Since $x^4 + y^4 = z^2$ has no positive integer solutions, then it follows that $x^4 + y^4 = z^4$ has no positive integer solutions. If $x^4 + y^4 = z^4$ did have a solution (x, y, z), then (x, y, z^2) would be a solution to $x^4 + y^4 = z^2$, which we now know is impossible.

This is as far as we can go with Fermat's Last Theorem. The proof for $n = 3$ is not too difficult, although we would need more machinery than we have at hand to tackle it. Some other special cases are also reasonably tractable. However, as we mentioned before, the general proof uses mathematics considerably beyond what you would encounter in any undergraduate (and most graduate) math courses.

6.6.   Fermat's Little Theorem

The *unit's digit* of a numeral is the digit in the "one's place"; that is, the digit at the right end of the numeral. It turns out that the unit's digits exhibit some interesting patterns.

1. *Complete the following table.*

Unit's digit of n	0	1	2	3	4	5	6	7	8	9
Unit's digit of n^2										
Unit's digit of n^3										
Unit's digit of n^4										
Unit's digit of n^5										

2. *What do you notice about the unit's digit of n^5?*

3. *What would you predict about the unit's digit of n^6? of n^7? of n^{43}?*

4. *What is the unit's digit of 139^{23}?*

5. *Is it possible for the numeral 3***7 (where the * indicates missing digits) to be a perfect square? Explain.*

6. *What is the largest value of $n \leq 50$ such that $3259^n - 2347^n$ is a multiple of 10?*

7. *Refer to the table above and state a theorem that allows us to determine the unit's digit of n^p for any positive integer p. In particular, your theorem should give a criterion for determining when the unit's digit of n^p and the unit's digit of n^q are the same.*

To prove this theorem, note that we need only consider the unit's digit of n^5.

8. *Why?*

9. *Restate the theorem about the unit's digit of n^5 as a theorem about the unit's digit of $n^5 - n$.*

10. *State an equivalent criterion about $n^5 - n$.*

11. *Prove the statement in question 10.*

6.6.1 Fermat's Little Theorem

12. *Prove that $n^2 - n$ is divisible by 2.*

13. *Prove that $n^3 - n$ is divisible by 3.*

14. *Is $n^5 - n$ divisible by 5? How do you know?*

Based on your observations above, you might be tempted to claim that $n^k - n$ is always divisible by k, for every positive integer k.

15. Do you think this theorem is true for all k? Can you find a counterexample?

16. If it is not true for all k, then for which k is it true? Complete the following theorem.

Theorem 6.7   If k _____ , then $n^k - n$ is divisible by k, for all n.

Note: Theorem 6.7 is known as *Fermat's Little Theorem*. It is not to be confused with Fermat's Last Theorem, which we mentioned in the last section.

While we can prove Fermat's Little Theorem for special cases, as in questions 12–14, we won't prove the general result at this time.

6.7.   Additional Questions

1.   Let P, Q, and R be points (in that order) on a line. Let $PQ = x$ and $QR = y$. Draw a semicircle with PR as diameter. Show that the radius of the semicircle is $A(x, y)$ and that the perpendicular distance from Q to the circle is $G(x, y)$. Does this prove the AGM Inequality?

2.   Use the AGM Inequality to find the minimum perimeter of a rectangle whose area is 144.

3.   Prove that if x, y, $z > 0$, then $\frac{x}{y} + \frac{y}{z} + \frac{z}{x} \geq 3$. Generalize.

4.   Prove that if x, y, $z > 0$, then $x^3 + y^3 + z^3 \geq 3xyz$. Generalize.

5.   Suppose x, y, $z > 0$ and $x + y + z = 6$. Show that:

(a) $\dfrac{1}{x} + \dfrac{1}{y} + \dfrac{1}{z} \geq \dfrac{3}{2}$ (b) $\left(x + \dfrac{1}{y}\right)^2 + \left(y + \dfrac{1}{z}\right)^2 + \left(z + \dfrac{1}{x}\right)^2 \geq \dfrac{75}{4}$

6.   Let $ABCD$ be a trapezoid with bases $AB = x$ and $CD = y$.
(a) A line drawn parallel to the bases divides the trapezoid into two trapezoids of equal area. Show that the length of this line is $R(x, y)$.
(b) The diagonals of the trapezoid intersect at F. Show that the length of the line through F parallel to the bases is $H(x, y)$.

7.   Prove that if $n \geq 2$, then $2^n \geq 1 + n\sqrt{2^{n-1}}$. [*Hint:* Use the AGM Inequality on the set of numbers $1, 2, 4, 8, \ldots, 2^{n-1}$.]

8.   Show that $1000!$ ends in 249 zeroes.

9. The goal of this problem is to prove that $\binom{2n}{n}$ is always an even integer.

(a) Show that the exponent of 2 in $(2n)!$ is given by $T = \sum_{k=1}^{\infty} \left[\frac{2n}{2^k} \right]$.

(b) Show that the exponent of 2 in $n!$ is given by $S = \sum_{k=1}^{\infty} \left[\frac{n}{2^k} \right]$ and prove that $S < n$.

(c) Show that $T - 2S = n - S$ and argue that $\binom{2n}{n}$ is always an even integer.

10. Hypothesize and prove a formula for $S_n = \sum_{k=1}^{n^2-1} \left[\sqrt{k} \right]$.

11. Prove that for any positive integer n,

$$\left[\frac{n}{3} \right] + \left[\frac{n+2}{6} \right] + \left[\frac{n+4}{6} \right] = \left[\frac{n}{2} \right] + \left[\frac{n+3}{6} \right].$$

[*Hint:* Let $n = 6q + r$, where $0 \le r < 6$. Consider six cases, one for each possible value of r.]

12. Let x be a real number and n be a positive integer. Show that $[nx] - n[x] \le n - 1$. [*Hint:* Let $x = a_0 + \frac{a_1}{n} + \frac{a_2}{n^2} + \cdots$, where all a_j's are integers and $0 \le a_j \le n - 1$, for $j = 1, 2, 3, \ldots$.]

13. (a) Show that $(2 + \sqrt{3})^n + (2 - \sqrt{3})^n$ is an even integer for all n.
(b) Argue that $[(2 - \sqrt{3})^n] = 0$.
(c) Prove that if x and y are not integers but $x + y$ is an integer, then $[x + y] = [x] + [y] + 1$.
(d) Using the above results, show that $[(2 + \sqrt{3})^n]$ is an odd integer for all n.

14. In a certain city, street parking is prohibited on the second and fourth Tuesdays of each month, so that the streets may be cleaned. Let x be the date on which a Tuesday falls. Determine a function f such that $f(x)$ tells which Tuesday it is. For example, if the 7th day of the month is a Tuesday, it would have to be the first Tuesday; thus, $f(7) = 1$. Similarly, $f(8) = 2$, $f(17) = 3$, and $f(25) = 4$.

15. Let $v_n = (u_n)^2$. Derive an expression for Δv_n in terms of Δu_n. Is your answer the same as the corresponding formula for derivatives?

16. Let

$$\binom{n}{r} = \frac{n!}{r!(n-r)!}.$$

For fixed r, determine the first difference of this sequence with respect to n.

17. Use the results of question 16 to show that

$$\sum_{k=1}^{n} \binom{k}{r} = \binom{n+1}{r+1}.$$

18. Evaluate $\displaystyle\sum_{k=1}^{n} \frac{1}{k(k+1)}$.

19. (a) Determine all sequences for which $\Delta u_n = u_n$.
(b) Determine all sequences for which $\Delta u_n = k u_n$, where k is a constant.

20. Marbles are sold in two packets of different size—one containing x marbles and the other containing y marbles. An order for a total of 207 marbles is to be sent in 17 packets of the first kind and 7 packets of the second kind. Determine x and y.

21. Find the smallest positive multiple of 11 that leaves a remainder of 1 when divided by 2, 3, 5, and 7.

22. Prove that the congruences $x \equiv a \bmod m$ and $x \equiv b \bmod n$ have a solution iff $a - b$ is divisible by $\gcd(m, n)$.

23. Prove that if $x \equiv a \bmod m$, then either $x \equiv a \bmod 2m$ or $x \equiv (a + m) \bmod 2m$.

24. Determine all Pythagorean triplets (not necessarily primitive) in which one side is 40.

25. Show that if (x, y, z) is a primitive Pythagorean triplet, then $x + y$ is congruent to either 1 or 7 mod 8.

26. Show that for any odd integer $n \geq 2$, $n^4 - 4$ and $4n^2$ can be the legs of a right triangle. How long is the hypotenuse?

27. A circle is inscribed in a right triangle ABC as shown below.
(a) Determine the area of the triangle in terms of x and y.
(b) Show that the sum of the areas of the three smaller triangles is $\frac{r(x + y + z)}{2}$.

(c) By equating the answers to (a) and (b), show that $r = \frac{xy}{x+y+z}$.

(d) Using the fact that (x, y, z) is a Pythagorean triplet, show that r is an integer.

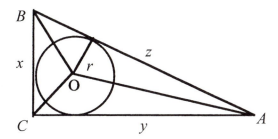

28. (a) Prove that one entry in a Pythagorean triplet is always divisible by 4.

(b) Prove that one entry in a Pythagorean triplet is always divisible by 3. [*Hint:* Consider two cases. In the first case, assume that either s or t is divisible by 3. In the second case, assume that neither s nor t is divisible by 3. Then show that s^2 and t^2 are both congruent to 1 mod 3.]

(c) Prove that one entry in a Pythagorean triplet is always divisible by 5.

29. Show that $x^2 + y^2 = z^3$ has infinitely many positive integer solutions. [*Hint:* Let $x = n(n^2 - 3)$, $y = 3n^2 - 1$, for any $n > 3$.]

30. Prove that Theorem 6.7 (Fermat's Little Theorem) is really a special case of Theorem 2.4 (Euler-Fermat Theorem).

31. In a triangle ABC, draw BD perpendicular to AC. Determine the smallest possible length of BD if $BD, AB, AC,$ and BC are consecutive integers (in that order). [*Hint:* Show that $(CD)^2 - 3(AD)^2 = 6$.]

32. What is the unit's digit of

$$(\dots ((7)^7)^7)^{\dots 7},$$

where there are 1000 sevens in the expression?

Chapter 7
Topics in Algebra

If you decide to major in mathematics, one of the courses you will likely be required to take is abstract algebra. Many of the ideas in such a course are generalizations of the things we've talked about in this book. In this chapter, we'll investigate a few of these generalizations. Although a typical algebra course would contain much more than we can cover here, this chapter should give you a good idea of what to expect if you continue your mathematics career.

7.1. Rings

At various points throughout the first six chapters of this book, we have studied sets of numbers such as the integers and rationals, and sets of other mathematical objects such as polynomials. In all cases, we defined arithmetic operations—addition, subtraction, multiplication, and division—on these sets. We then asked whether the sets were closed under these operations.

Depending on the answers to the closure questions, we then explored other properties. In particular, for sets that are closed under addition, subtraction, and multiplication, but not division (e.g., the integers), we had the Division Algorithm to tell us what happens when we divide. We also could talk about *primes* (which in the case of polynomials, we called *irreducibles*) and related issues such as unique factorization.

In some instances, the situation was more complicated than we might have expected. For example, we were initially led to believe that the set $Z[\sqrt{2}] = \{a + b\sqrt{2},$ where a and b are integers$\}$ did not have unique factorization since $14 = 2 \times 7 = (4 - \sqrt{2})(4 + \sqrt{2})$. However, this conclusion was premature, since we had not yet defined what the primes were. It turns out that neither 2 nor 7 nor $4 - \sqrt{2}$ nor $4 + \sqrt{2}$ are prime, since all can be expressed as the product of two elements of S; for example, $7 = (3 - \sqrt{2})(3 + \sqrt{2})$.

We also were tempted to claim that the set of polynomials with integer coefficients did not have unique factorization since $4x^2 - 16 = (2x - 4)(2x + 4) = (x - 2)(2x + 8)$, where all factors are irreducible. However, if we agree to first factor out the greatest common divisor of the coefficients, then we do get unique factorization, except for factors of ± 1.

None of the questions about factoring and primes make sense for the rational numbers or the real numbers, which are closed under division. However, we can pursue other questions relevant to these sets.

In this chapter, we are going to investigate in more generality many of the questions that we have been exploring in specific instances so far. In the first two sections, we'll consider sets with structures like the integers. Such sets are called *rings*. Then, we'll consider sets with structures like the rationals. Such sets are called *fields*.

We begin with a definition.

Definition   Let S be a set on which two operations, called *addition* and *multiplication*, are defined. S is called a *ring* if the following properties hold.

1. If a and b are elements of S, then $a + b$ is an element of S.
2. $a + b = b + a$, for all a and b in S.
3. $a + (b + c) = (a + b) + c$, for all a, b, and c in S.
4. There exists an element θ in S such that $a + \theta = a$, for all a in S.
5. For each element a in S, there exists an element $(-a)$ such that $a + (-a) = \theta$.
6. If a and b are elements of S, then ab is an element of S.
7. $(ab)c = a(bc)$, for all a, b, and c in S.
8. $a(b + c) = ab + ac$ and $(b + c)a = ba + ca$, for all a, b, and c in S.

Properties (1) and (6) are the closure properties under addition and multiplication, respectively. Property (2) says that addition is *commutative*; properties (3) and (7) say that addition and multiplication are *associative*. Property (8) is the *distributive* property of multiplication over addition. The element θ in property (4) is called the *additive identity*. Property (5) says that every element has an *additive inverse*.

Notice that we do not require multiplication to be commutative. If, in addition to the properties above, multiplication were commutative, S would be called a *commutative ring*.

Some rings have a multiplicative identity element ε with the property that $\varepsilon a = a\varepsilon = a$ for all a in the ring. Note that since multiplication is not commutative, we need to specify that the identity element "works both ways" (i.e., multiplying on the left and on the right).

Elements of a ring need not have multiplicative inverses. First of all, the concept of multiplicative inverse would not make sense if there were no multiplicative identity element. Second, if every element did have a multiplicative inverse, then the set would be, in essence, closed under division. This would make the set a field.

Finally, note that sets such as the rationals are rings, even though they have properties beyond those listed in the definition above.

1. *Argue that the second part of property (8) is redundant in a commutative ring.*

2. *How would you define "subtraction" in a ring? Are rings closed under subtraction? Explain.*

3. *Which of the following are rings? For each that is not a ring, state which property fails. (There may be more than one.) Also, identify the additive identity element in those that are rings.*
    - (i) *{even integers}*
   - (ii) *{odd integers}*
  - (iii) *{nonnegative integers}*
   - (iv) *{polynomials with integer coefficients}*
    - (v) *{second-degree polynomials with integer coefficients}*
   - (vi) *{polynomials with rational coefficients}*
  - (vii) *{polynomials with coefficients in Z_4}*
 - (viii) *$\{a + b\sqrt[3]{2}$, where a and b are integers\}*
   - (ix) *$\{a + b\sqrt[3]{2} + c\sqrt[3]{4}$, where a, b and c are integers\}*
    - (x) *the set $S = \{(x, y)$, where x and y are real numbers\}, where we define addition by $(x_1, y_1) + (x_2, y_2) = (x_1 + x_2, y_1 + y_2)$ and multiplication by $(x_1, y_1) \cdot (x_2, y_2) = (x_1x_2, y_1y_2)$.*

4. *Which of the rings in question 3 have a multiplicative identity element?*

To determine whether a given set is a ring under the given operations, all we need to know is the "sum" and "product" of each pair of elements in the set. We can give these results in table form. Let $S = \{\theta, a, b, c\}$. Addition and multiplication are defined by the following tables.

+	θ	a	b	c
θ	θ	a	b	c
a	a	θ	c	b
b	b	c	θ	a
c	c	b	a	θ

·	θ	a	b	c
θ	θ	θ	θ	θ
a	θ	a	b	c
b	θ	a	b	c
c	θ	θ	θ	θ

To show that this is a ring, we have to show that all the properties in the definition hold. The closure properties are obvious since there are no entries in the tables that are not elements of S. There is an additive identity element θ since $\theta + \theta = \theta$, $a + \theta = a$, $b + \theta = b$, and $c + \theta = c$.

5. What are the additive inverses of each of the elements?

Verifying the commutative, associative, and distributive properties is more difficult since we must consider all possible pairs or triples of elements. The commutativity of addition is fairly easy to spot since there is a "symmetry" in the addition table—the first row is identical to the first column, the second row is identical to the second column, and so on.

To show that addition is associative, we see, for example, that $(a + b) + c = c + c = \theta$ and $a + (b + c) = a + a = \theta$. Also, $(a + a) + c = \theta + c = c$ and $a + (a + c) = a + b = c$. These are just two possibilities; we'd have to check all the others. Then we'd have to repeat the process for the associative property of multiplication and the distributive properties.

6. Check a few cases for each of the properties to make sure you understand what the properties mean.

7. Is multiplication commutative in this ring? Is there a multiplicative identity element?

It is important to remember to "play by the rules" and not be misled by old, ingrained algebraic habits. For example, we may write $a + a = 2a$, but we cannot interpret $2a$ as "2 times a" since 2 may not even be an element of the ring. Also, using the distributive properties, we have $(a + b)(a + b) = a^2 + ab + ba + b^2$. This, in turn, is *not* the same as $a^2 + 2ab + b^2$, unless the ring is commutative. Finally, we may have $ab = ac$, where $a \neq \theta$, but this does not imply $b = c$.

Nonetheless, if we are careful, we can prove a variety of facts about rings. For example, consider the following lemma.

Lemma 7.1   Let a, b, and c be elements of a ring. If $a + b = a + c$, then $b = c$.

8. *Justify each of the following statements in the proof of Lemma 7.1.*

   (i) Since $a + b = a + c$, then $(-a) + (a + b) = (-a) + (a + c)$.
   (ii) $((-a) + a) + b = ((-a) + a) + c$.
   (iii) $b = c$.
9. *Prove that $-(-a) = a$ for all a in the ring.*

The next theorem gives a list of properties of rings that can be proven from the basic definition.

Theorem 7.1   Let a, b, and c be elements of a ring S. Then,

   (i) $a\theta = \theta a = \theta$.
   (ii) $a(-b) = (-a)b = -ab$.
   (iii) $(-a)(-b) = ab$.
   (iv) If S has a multiplicative identity element ε, then $(-\varepsilon)a = -a$.

To prove (i), observe that $\theta + a\theta = a\theta = a(\theta + \theta) = a\theta + a\theta$.

10. *Justify each equality in the statement above.*

11. *Why can we conclude that $a\theta = \theta$?*

12. *Prove that $\theta a = \theta$.*

To prove (ii), observe that $\theta = a\theta = a((-b) + b) = a(-b) + ab$.

13. *Justify each equality in the statement above.*

14. *Complete the proof.*

15. *Prove (iii) and (iv). [Hint: Use (ii).]*

7.1.1. Integral Domains

Given two integers x and y, we know that if $xy = 0$, then either $x = 0$ or $y = 0$. Since the integers are a ring, it makes sense to ask whether this property holds for all rings.

We have already seen that the answer is no. For example, in Z_6, we have $2(3) = 0$; that is, the product of two nonzero elements is 0. The ring displayed by table following question 4 is another example; there, $ca = \theta$ and neither c nor a is the additive identity θ.

We say that an element c in a ring is a *divisor of zero* if there exists an element b in the ring such that $bc = \theta$. We can distinguish between those rings that have divisors of zero and those that don't.

Definition: A commutative ring with a nonzero multiplicative identity element is said to be an *integral domain* if it has no zero divisors.

Hence, the integers (conveniently) are an integral domain; Z_6 is not. We need the ring to be commutative; otherwise, we could have $bc = \theta$ but $cb \neq \theta$. So we'd have to be careful about which side we were multiplying on.

16. For which values of n is Z_n an integral domain?

Proving that a ring is an integral domain is a straightforward, but not always trivial, process. Consider the set $Z[\sqrt{2}] = \{a + b\sqrt{2}$, where a and b are integers\}. We know that $Z[\sqrt{2}]$ is a commutative ring with multiplicative identity element 1. To show that $Z[\sqrt{2}]$ is also an integral domain, let $x = a + b\sqrt{2}$ and $y = c + d\sqrt{2}$ be two elements of S.

17. Show that $xy = 0$ iff $ac + 2bd = 0$ and $ad + bc = 0$. Then show that this implies $b(2d^2 - c^2) = 0$.

18. Since the integers are an integral domain, then either $b = 0$ or $2d^2 = c^2$. Argue that the latter is impossible. Thus, $b = 0$.

19. Now argue that either $a = 0$ or $c = 0$. In either case, show that either $x = 0$ or $y = 0$. This completes the proof.

20. Prove the following theorem.

Theorem 7.2:   Let a, b, and c be elements of an integral domain. If $a \neq \theta$ and $ab = ac$, then $b = c$.

Keep in mind that this theorem is not trivial because elements of a ring do not necessarily have multiplicative inverses. Suppose they did. Then, letting a^{-1} denote the multiplicative inverse of a (meaning that $a^{-1}(a) = \varepsilon$, where ε is the multiplicative identity element), we'd have $a^{-1}(ab) = a^{-1}(ac)$. Using the associative property of multiplication and the definition of the multiplicative inverse, we conclude that $b = c$. Theorem 7.2 tells us that, even though there are no multiplicative inverses, we can still do this "multiplicative cancellation." The next theorem tells us something about rings of polynomials.

Theorem 7.3   Let D be an integral domain and let

$$D[x] = \{\text{polynomials with coefficients in } D\}.$$

Then $D[x]$ is an integral domain.

21. To prove Theorem 7.3, we need to show three things. First show that $D[x]$ is commutative. Then show that $D[x]$ has a nonzero multiplicative identity element.

22. Finally, choose two arbitrary nonzero elements of $D[x]$, say, $p(x) = a_n x^n + a_{n-1} x^{n-1} + \cdots + a_0$ and $q(x) = b_m x^m + b_{m-1} x^{m-1} + \cdots + b_0$. Show that $p(x)q(x) \neq 0$.

7.2. Unique Factorization

In this section, we'll attempt to settle the questions raised earlier about unique factorization. First, we need a definition.

Definition Let S be a commutative ring with multiplicative identity ε. A nonzero element u of S is said to be a *unit* of S if there exists an element v in S such that $uv = \varepsilon$.

In other words, the units of a ring are those elements with multiplicative inverses in the ring. For example, the only units in the set of integers are $+1$ and -1. However, in the rational numbers, all nonzero elements are units since they all have rational multiplicative inverses (or reciprocals).

To determine the units in Z_n, it is helpful to look at a multiplication table. Here's the table for Z_4:

$\cdot$	0	1	2	3
0	0	0	0	0
1	0	1	2	3
2	0	2	0	2
3	0	3	2	1

Since the multiplicative identity element is 1, the units are those elements whose rows (or columns) have a 1 in them. Hence, the units are 1 and 3. Note that each of the units is its own inverse.

1. What are the units in Z_5? in Z_6? in Z_8? in Z_{10}? What are the inverses of each of the units?

2. Complete and prove the following theorem. For the proof, you might wish to refer to Theorems 6.2 and 6.3.

Theorem 7.4 u is a unit of Z_n iff _____.

Determining the units in other rings is more complicated. For instance, consider the set $Z[\sqrt{2}] = \{a + b\sqrt{2}$, where a and b are integers$\}$. The multiplicative identity is 1; hence, $u = a + b\sqrt{2}$ is a unit iff there exists $v = c + d\sqrt{2}$ such that $uv = 1$.

3. *Show that this implies* $c + d\sqrt{2} = \dfrac{a - b\sqrt{2}}{a^2 - 2b^2}.$

4. *Argue that those elements of S for which* $a^2 - 2b^2 = \pm 1$ *are units.*

It is also true that these are the only units of S. (See Additional Questions in Section 7.4 for proof.) Determining all the units—all the solutions to the equation above—is not easy. There is some symmetry; specifically, if $a + b\sqrt{2}$ is a unit, then $a - b\sqrt{2}$, $-a + b\sqrt{2}$, and $-a - b\sqrt{2}$ are also units. So, we get four for the price of one.

5. *Find 12 units (3 sets of 4) in S.*
6. (i) *Let* $Z[i] = \{a + bi$, *where a and b are integers*$\}$. *Show that the only units in this ring are* 1, -1, i, *and* $-i$.
   (ii) *Determine all the units in the ring*

   $$Z[\sqrt{-5}] = \{a + b\sqrt{5}i, \text{ where } a \text{ and } b \text{ are integers}\}.$$

7. *Determine all units in each of the following.*
   (i) $Z[x] = \{$*polynomials with integer coefficients*$\}$
   (ii) $Q[x] = \{$*polynomials with rational coefficients*$\}$

Definition   In a given ring, we say that a is an *associate* of b, where a and b are not equal to θ, if there exists a unit u such that $a = ub$.

For example, since the only units in the set of integers are ± 1, then every integer n has an associate $-n$.

8. *Show that if a is an associate of b, then b is an associate of a. Hence, we can just say that "a and b are associates."*

9. *List all pairs of associates in* Z_6.
10. (i) *What are all the associates of* $3 + 2i$ *in* $Z[i]$?
    (ii) *What are all the associates of* $2 + 3\sqrt{5}i$ *in* $Z[\sqrt{-5}]$?
11. (i) *What are the associates of* $2x + 5$ *in* $Z[x]$?
    (ii) *What are the associates of* $2x + 5$ *in* $Q[x]$?

We need one last definition before we can talk about unique factorization.

Definition   An element c of an integral domain D is said to be *irreducible* if c is not a unit and if $c = ab$, where a and b are elements of D, then either a or b is a unit.

In other words, the irreducible elements are those that cannot be expressed as the product of two other elements, unless one of those elements is a unit. So, irreducible elements are the nonunits that are divisible only by their associates and the units.

In **Z**, the irreducible elements are the primes and their negatives—that is, the set $P = \{\pm 2, \pm 3, \pm 5, \pm 7, \ldots\}$. No element of P can be factored in a way in which neither factor is $+1$ or -1, the units of **Z**. Now we see why 1 is not considered prime.

In Z_6, the units are 1 and 5. None of the other elements in Z_6 are irreducible since $2 = 2(4)$, $3 = 3(3)$, and $4 = 4(4)$.

It is important that we specify the integral domain in which we are working. Elements that are irreducible in one domain may be reducible in another. In $Z[i]$, the units are 1, -1, i, and $-i$. The integer 2, which is irreducible in **Z**, is reducible in $Z[i]$ since $2 = (1 + i)(1 - i)$ and neither factor is a unit.

12. *Find three other prime integers that are reducible in Z[i]. Find three prime integers that are irreducible in Z[i].*

13. *Determine all irreducibles in Z_8.*

14. *Is $p(x) = 3x + 12$ irreducible in Z[x]? in Q[x]?*

Definition   An integral domain D is said to be a *unique factorization domain* if both of the following hold.

  (i) Each element of D that is not a unit can be written as a product of irreducibles,

AND

  (ii) if $p_1 p_2 \ldots p_m$ and $q_1 q_2 \ldots q_n$ are two distinct factorizations of the same element x, where all factors are irreducible, then $m = n$ and the factors can be rearranged so that p_i is an associate of q_i for all i.

Part (ii) gives a precise interpretation to the unique factorization property to which we have often referred in previous chapters.

The Fundamental Theorem of Arithmetic (Theorem 1.7) ensures that the integers **Z** are a unique factorization domain (henceforth abbreviated UFD). Even though there are multiple factorizations of any integer, these factorizations are essentially the same, in the sense given by (ii) above. For example, $15 = 3 \times 5 = (-3) \times (-5)$, but 3 is an associate of -3 and 5 is an associate of -5.

In $Q[x]$, the units are all the rational numbers—that is, nonzero constant polynomials. All linear polynomials are irreducible in $Q[x]$. So, the two factorizations $(x + 2)(4x - 8)$ and $(2x + 4)(2x - 4)$ of $4x^2 - 16$ do not violate the UFD definition since $x + 2$ is an associate of $2x + 4$ and $4x - 8$ is an associate of $2x - 4$. This example is evidence, but not proof, that $Q[x]$ is a UFD.

In $Z[\sqrt{2}]$, we have shown that the units are any elements of the form $a + b\sqrt{2}$ for which $a^2 - 2b^2 = \pm1$. Therefore, ±1, $\pm1 \pm \sqrt{2}$, and $\pm3 \pm 2\sqrt{2}$, among others, are units. Now, $2 = \sqrt{2}\sqrt{2} = (2 - \sqrt{2})(2 + \sqrt{2})$, apparently two distinct factorizations of 2 as the product of irreducibles. However, $\sqrt{2} = (2 - \sqrt{2})(1 + \sqrt{2})$; hence, $\sqrt{2}$ is an associate of $2 - \sqrt{2}$. Similarly, $\sqrt{2} = (2 + \sqrt{2})(-1 + \sqrt{2})$ so $\sqrt{2}$ is also an associate of $2 + \sqrt{2}$. Thus, the two factorizations are essentially the same. Again this is evidence, but not proof, that $Z[\sqrt{2}]$ is a UFD.

15. *Is Z_6 a UFD?*

16. *In $Z[i]$, we have $5 = (1 + 2i)(1 - 2i) = (2 + i)(2 - i)$, where all factors are irreducible. Reconcile this with the UFD properties.*

Proving that a given integral domain is a UFD is not easy. It turns out that we can relate unique factorization to the existence of a "division algorithm." First, we need to clarify what we mean by this.

Definition An integral domain D is said to be a *Euclidean domain* if there exists a function d such that:

  (i) $d(a)$ is a nonnegative integer for all nonzero elements of D;
  (ii) $d(a) \le d(ab)$, for all nonzero a and b in D; and
  (iii) for any a and b in D, with $b \ne 0$, there exist elements q and r in D
       such that $a = bq + r$, where $r = 0$ or $d(r) < d(b)$.

This definition is reminiscent of the Division Algorithm (Theorem 1.2) encountered in Chapter 1, but we need to reconcile a few things. Theorem 1.2 says that for any integers a and b, with $b > 0$, there exist unique integers q and r such that $a = bq + r$, where $0 \le r < b$.

17. *Convince yourself that if $b < 0$, we could modify the last phrase to "where $0 \le |r| < |b|$," as long as we are willing to forgo the "uniqueness."*

It follows that (iii) of the definition is satisfied if we take $d(x) = |x|$. Now we have to check whether this choice of d satisfies (i) and (ii).

18. Do so.

Hence, **Z** is a Euclidean domain. Notice that (iii) of the definition does not require q and r to be unique. Although sometimes they are unique, there are examples of Euclidean domains in which they aren't.

In Chapter 3, we investigated the existence of a division algorithm for polynomials (Theorem 3.1). There we said that for any two polynomials f and g, there exist polynomials q and r such that $f(x) = g(x)q(x) + r(x)$, where degree of $r(x)$ < degree of $g(x)$.

19. Show that the function $d[p(x)] = degree$ of $p(x)$ satisfies the conditions of the definition of a Euclidean domain.

Hence, $Q[x]$ is a Euclidean domain. Before we look at any more examples, let's state a theorem that relates Euclidean domains to unique factorization domains. The proof is beyond the scope of this course.

Theorem 7.5 If D is a Euclidean domain, then it is a UFD.

Thus, to determine whether a given integral domain has unique factorization, it may be easier to show that it is Euclidean, rather than try to verify the UFD properties directly.

7.2.1. Gaussian Integers

Let $Z[i] = \{a + bi \mid a$ and b are integers$\}$, which is a subset of the set of complex numbers. We call $Z[i]$ the set of *complex integers* or *Gaussian integers*. We know that $Z[i]$ is a commutative ring with identity element 1.

20. Show that $Z[i]$ is an integral domain.

The question we'd like to answer is whether $Z[i]$ is a UFD. To show that it is, we'll show that it is Euclidean. First, let's look at an example of what happens when we divide two elements of $Z[i]$.

21. Show that $\dfrac{8 - 5i}{1 - 2i} = \dfrac{18 + 11i}{5} = \dfrac{18}{5} + \dfrac{11}{5}i.$

Note that the quotient is not a complex integer; hence, $Z[i]$ is not closed under division.

So now the question becomes: Is there a function d that satisfies the definition of a Euclidean domain? In other words, is there a function d such that:

(i) $d(z)$ is a nonnegative integer for all $z \neq 0$;

(ii) $d(z) < d(zw)$ for all nonzero z and w; and

(iii) given two Gaussian integers $z = a + bi$ and $w = c + di$, there exist Gaussian integers q and r such that $z = qw + r$, where either $r = 0$ or $d(r) < d(w)$?

For clarity, let's use the term *quotient* to mean the exact, possibly nonintegral result when two integers are divided. We'll use the term *integer quotient* to represent the "q" term in the definition of Euclidean domain. For ordinary integers, the integer quotient is obtained by rounding the quotient down to the nearest integer. (For example, when we divide 13 by 6, the integer quotient is 2, which is the largest integer less than $\frac{13}{6}$.) This ensures that the remainder will be nonnegative and smaller than the divisor.

For the complex integers, it might seem natural to define the integer quotient as the complex integer obtained by rounding the real and imaginary parts of the quotient down to the nearest integer. In the example above, the integer quotient would be $3 + 2i$.

22. Show that the corresponding remainder is $1 - i$.

The next question is whether the remainder is smaller than the divisor. First, we need to define what we mean by "smaller." In other words, we need to define the function d. One possibility is to use the *magnitude*; that is, let $d(z) = \|z\|$.

23. Show that $d(z) = \|z\|$ does not satisfy (i) above.

An alternative might be to let $d(z) = \|z\|^2$. This now satisfies (i) since $\|z\|^2 = a^2 + b^2$, which is a nonnegative integer as long as $z \neq 0$.

24. Show that (ii) is satisfied by $d(z) = \|z\|^2$. [Hint: Remember that $\|zw\| = \|z\|\,\|w\|.]$

25. Show that the remainder obtained in question 22 is less than the divisor, where "less than" is defined by the function $d(z) = \|z\|^2$.

It remains to show that (iii) can be satisfied for every choice of z and w. We appeal to a geometric argument. The actual quotient $\frac{z}{w}$ corresponds to the point P on the complex plane, as shown in the graph below. The vertices of $ABCD$ are Gaussian integers obtained by rounding the real and imaginary parts of the quotient either up or down to the nearest integer. When we round the real and imaginary parts down to the nearest integer, as in question 21, we moved to point A. Hence, for that example, A is the point corresponding to q.

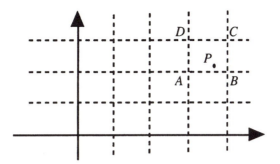

Since $z = qw + r$, then $\dfrac{z}{w} = q + \dfrac{r}{w}$.

26. *Argue that* $\left\|\dfrac{z}{w} - q\right\| = $ *length of PA* $ = \left\|\dfrac{r}{w}\right\| = \dfrac{\|r\|}{\|w\|}$.

Consequently, $d(r) < d(w)$ iff the length of $PA < 1$. Clearly, this need not be the case in general. It is possible that P will be more than one unit from A but still inside the square $ABCD$.

27. *Argue that one of the vertices of the square must be less than 1 unit from P.*

Therefore, we could take as the integer quotient q any vertex that is within 1 unit of the quotient $\frac{z}{w}$. We may have to round the real and imaginary parts up or down to accomplish this.

Finally, the division algorithm for integers assures us that the quotient and remainder are unique. Although this is not required in the definition of Euclidean domain, let's see if it is true here. In other words, in the example in question 21, is $3 + 2i$ the only quotient that would produce a remainder "less than" the divisor?

28. *Show that we could have also chosen the quotient to be* $4 + 2i$.

29. *There are two other possibilities—the remaining two vertices of the unit square surrounding the actual quotient. Check each to see if they are acceptable as well.*

30. *Prove that there are always at least two choices; that is, show that no matter where the actual quotient is located, there are always at least two vertices that are less than one unit away.*

31. *Geometrically, describe an example in which there would be*
    (i) three choices.

(ii) four choices.

Since we have now shown that the Gaussian integers are a Euclidean domain, it follows from Theorem 7.5 that they are a unique factorization domain as well. The logical next step is to identify the irreducible elements of $Z[i]$. The answer is given by the following theorem.

Theorem 7.6   Let $z = a + bi$, where $a \neq 0$ and $b \neq 0$. Then z is irreducible in $Z[i]$ iff $d(z) = \|z\|^2 = a^2 + b^2$ is prime. If $b = 0$, then z is irreducible iff $|a|$ is a prime that is congruent to 3 mod 4. Similarly, if $a = 0$, then z is irreducible iff $|b|$ is a prime that is congruent to 3 mod 4.

Part of the proof of this theorem—the "if" part—is straightforward; the rest is beyond the scope of this course. Note that in the case where $b = 0$, then $z = a$; if $|a|$ is composite, then z is surely reducible. A similar argument applies if $a = 0$.

32. Prove that units in $Z[i]$ are the only elements for which $d(z) = 1$.

33. Prove that if $d(z)$ is prime, then z is irreducible.

34. Which of the following elements of $Z[i]$ are irreducible?
    (i) $2 + 3i$ (ii) $3 + 4i$ (iii) 11
    (iv) $17i$ (v) 29

35. Express each of the reducible elements above as a product of irreducibles. Is your answer necessarily unique?

The technique we've used here can be adapted to show that other sets, such as $Z[\sqrt{2}]$, are Euclidean and, hence, are UFDs. It would be a mistake, however, to assume that all sets of the form $Z[\sqrt{k}]$ are UFDs.
    For example, let $Z[\sqrt{-5}] = \{a + b\sqrt{5}i$, where a and b are integers$\}$. It is easy to show that $Z[\sqrt{-5}]$ is an integral domain.

36. Show that u is a unit of $Z[\sqrt{-5}]$ iff $d(u) = a^2 + 5b^2 = 1$.

37. Show that the only units in $Z[\sqrt{-5}]$ are $u = \pm 1$.

38. Show that $46 = 2 \times 23 = (1 + 3\sqrt{5}i)(1 - 3\sqrt{5}i)$.

39. Show that 2 and 23 are irreducible. [Hint: If 2 were reducible, then $2 = zw$, where neither z nor w is a unit. This implies $4 = d(2) = d(z)d(w)$. What does this say about $d(z)$ and $d(w)$?]

A similar argument can be used to show that $1 + 3\sqrt{5}i$ and $1 - 3\sqrt{5}i$ are irreducible. Hence we have expressed 46 as the product of irreducibles in two distinct ways, in the sense that neither set of factors are associates of the others.

7.3. Fields

As we indicated earlier, sets such as the rationals, reals, and complex numbers, in addition to having the properties of a commutative ring with identity, are closed under division. This suggests the following definition.

Definition   A set F is said to be a *field* if F is a commutative ring with a nonzero multiplicative identity element ε and if, for every $x \neq 0$, there exists an element y such that $xy = \varepsilon$.

We say that y is the *multiplicative inverse* of x and we write $y = x^{-1}$. We think of the quotient $\frac{a}{b}$ as the product ab^{-1} (or $b^{-1}a$, since multiplication is commutative).

Put differently, a commutative ring with identity is a field if every element is a unit.

1. *Show that $Q[\sqrt{2}] = \{a + b\sqrt{2}$, where a and b are rational$\}$ is a field.*

2. *Let $Z_2[i] = \{a + bi$, where a and b are elements of $Z_2\}$. Create a multiplication table for $Z_2[i]$, assuming all arithmetic is done in Z_2. Is $Z_2[i]$ a field?*

3. *Prove that if F is a field, then F is an integral domain.*

The converse of question 3 is not necessarily true. In other words, there are integral domains—such as the integers—that are not fields. However, we can state the following theorem.

Theorem 7.7   If F is a finite integral domain, then F is a field.

To prove this theorem, let x be an arbitrary nonzero element of F. We need to show that there exists y such that $xy = \varepsilon$. If $x = \varepsilon$, then we are done; simply take $y = \varepsilon$. If $x \neq \varepsilon$, then we need to consider the sequence $\{x, x^2, x^3, \ldots\}$.

4. *Argue that there must be two entries in this sequence that are equal; that is, there exist distinct positive integers r and s such that $x^r = x^s$. Without loss of generality, we may assume $r > s$.*

5. *This implies $x^{r-s} = \varepsilon$. Why?*

6. *What is the multiplicative inverse of x?*

7. *Show that Z_3 and Z_5 are fields. What is the inverse of each element?*

8. *Complete and prove the following theorem.*

Theorem 7.8 Z_n is a field iff ⸻ .

Let's take a closer look at $Z_n = \{0, 1, 2, \ldots, n - 1\}$. We can think of these elements as disjoint subsets of the integers. In other words, the element "2" represents all the integers that are congruent to 2 mod n; the element "3" represents those integers congruent to 3 mod n, and so on. The Division Algorithm guarantees that all integers are represented by one of these subsets.

Arithmetic operations in Z_n can be interpreted in the context of these subsets. When we say, for instance, that in Z_5, $3 + 4 = 2$, we really mean that the sum of *any* number congruent to 3 mod 5 and *any* number congruent to 4 mod 5 is a number that is congruent to 2 mod 5.

The idea of partitioning a ring (like the integers) into disjoint subsets, where all elements in each subset are "congruent" to each other in some sense, can be generalized. Although a discussion of the most general case is beyond the scope of this text, we will look at another example.

Let $F[x] = \{$polynomials with coefficients in $F\}$, where F is some field, and let $p(x)$ be a given polynomial in $F[x]$.

9. Show that $F[x]$ is a commutative ring with identity.

We can partition the polynomials in $F[x]$ into disjoint subsets, one for each possible remainder when that polynomial is divided by $p(x)$. We'll use the notation $F[x]/\langle p(x) \rangle$ to represent this set of subsets. For purposes of this discussion, we put $\langle \rangle$ around $p(x)$ to avoid confusion with ordinary division of $F[x]$ by $p(x)$—here we want to divide and take the remainder.[1] By analogy, we could write Z_n as $Z/\langle n \rangle$. One difference is that, depending on the choice of F, $F[x]/\langle p(x) \rangle$ may contain infinitely many elements, whereas Z_n always contains a finite number of elements.

Let's look at an example. Suppose $F = \mathbf{Q}$ (the rational numbers) and $p(x) = x - c$, for some rational constant c.

10. Argue that $Q[x]/\langle x - c \rangle = \mathbf{Q}$; that is, the set of possible remainders is just the set of rational numbers.

11. Show that $f(x)$ and $g(x)$ belong to the same subset (i.e., element of $Q[x]/\langle x - c \rangle$) iff $f(c) = g(c)$.

Note that in this example, $F[x]/\langle p(x) \rangle$ is itself a field. The question we'd like to answer is whether that is always true; in other words, is $F[x]/\langle p(x) \rangle$ a field for every choice of field F and polynomial $p(x)$?

---

[1]In fact, $\langle p(x) \rangle$ represents the set of all polynomials of the form $f(x)p(x)$, where $f(x)$ is an element of $F[x]$. It is called the *ideal* generated by $p(x)$. The theory of ideals was first proposed in the mid-nineteenth century by Ernst Kummer, who was attempting to prove Fermat's Last Theorem. You can find out more about ideals in any abstract algebra text.

Let's try a more complicated example. Keep $F = \mathbf{Q}$, but let $p(x) = x^2 + 1$. Now $F[x]/\langle p(x)\rangle$ contains all possible remainders when a polynomial with rational coefficients is divided by $x^2 + 1$.

12. Argue that $Q[x]/\langle x^2 + 1\rangle = \{ax + b, \text{ where } a \text{ and } b \text{ are rational}\}$; that is, the set of first-degree polynomials with rational coefficients.

13. To which subset (i.e., element of $Q[x]/\langle x^2 + 1\rangle$) does each of the following polynomials belong?
   (i) $f(x) = x^3 + 2x + 1$ (ii) $g(x) = x^4 - 2x^2 + 3x - 5$

14. Determine any three other polynomials that belong to the same subset as $f(x) = x^3 + 2x + 1$.

15. What must be true about the difference between any two polynomials belonging to the same subset?

The next step is to define arithmetic operations within this set. We do it in much the same way that we did it in Z_n—that is, do ordinary arithmetic and then "reduce" the result to an element of the set. Addition in $Q[x]/\langle x^2 + 1\rangle$ is easy since the sum of two linear polynomials is (at worst) linear, so there is no "reduction" involved. Multiplication is a different story. For example, $(2x + 1)(3x - 2) = 6x^2 - x - 2 = 6(x^2 + 1) - x - 8$ in ordinary algebra. Hence, in $Q[x]/\langle x^2 + 1\rangle$, $(2x + 1)(3x - 2) = -x - 8$, the remainder after dividing the product by $x^2 + 1$.

16. Compute $(4x + 3)(2x + 5)$ in $Q[x]/\langle x^2 + 1\rangle$.

17. Show that $Q[x]/\langle x^2 + 1\rangle$ is a commutative ring with identity.

To determine whether $Q[x]/\langle x^2 + 1\rangle$ is a field, we have to determine whether each nonzero element has a multiplicative inverse. In other words, given $f(x) = ax + b$, does there exist $g(x) = cx + d$ such that $f(x)g(x) = 1$? (Remember, the multiplication is done in $Q[x]/\langle x^2 + 1\rangle$; otherwise the answer would clearly be no.)

18. Show that $f(x)g(x) = 1$ is equivalent to the system of equations $ad + bc = 0$, $bd - ac = 1$.

19. Solve this system for c and d in terms of a and b.

20. Why does the set in which the coefficients of the polynomial (in this case $\mathbf{Q}$) have to be a field? Would it suffice for it to be a ring?

21. Fill in the remaining details to show that $Q[x]/\langle x^2 + 1\rangle$ is indeed a field.

We now have two examples in which $F[x]/\langle p(x)\rangle$ is a field. This is not, however, strong enough evidence to conclude that $F[x]/\langle p(x)\rangle$ is always a field. Consider

$Q[x]/\langle x^2 - 1\rangle$. As before, let $f(x) = ax + b$ and $g(x) = cx + d$ and consider the equation $f(x)g(x) = 1$.

22. *Modify your calculations to questions 18 and 19 and show that*

$$c = \frac{-a}{a^2 - b^2} \text{ and } d = \frac{b}{a^2 - b^2}.$$

23. *There are some values of a and b for which c and d do not exist. Which ones? Which elements of $Q[x]/\langle x^2 - 1\rangle$ do not have multiplicative inverses?*

24. *Is $Q[x]/\langle x^2 - 1\rangle$ a field?*

25. *What is the distinction between $Q[x]/\langle x^2 - 1\rangle$ and $Q[x]/\langle x^2 + 1\rangle$ that makes one of them a field and the other just a commutative ring with identity? [Hint: Think about Z_n.]*

26. *Complete the following theorem.*

Theorem 7.9 Let F be a field and $p(x)$ be a polynomial. Then $F[x]/\langle p(x)\rangle$ is a field iff $p(x)$ is ——————— .

The proof makes use of an analogy to Theorems 6.2 and 6.3, which say that the Diophantine equation $ax + by = c$ has a solution iff c is divisible by $\gcd(a, b)$. In particular, if a and b are relatively prime, then $ax + by = c$ has a solution for every c. Taking this one step further, if b is prime, then $ax + by = 1$ has a solution for every a.

In terms of polynomials, we have the analogous statement that if $p(x)$ is irreducible, then for every $f(x)$, there exist polynomials $g(x)$ and $q(x)$ in $F[x]$ such that $f(x)g(x) + p(x)q(x) = 1$. This means that $f(x)g(x)$ is "congruent to 1 mod $p(x)$."

27. *Why does this prove Theorem 7.9?*

While there is a lot more we could say about rings and fields, it is best to leave that to a textbook on abstract algebra. Here we'll settle for a taste, one that we hope will whet your appetite to study further.

7.4. Additional Questions

1. Let R be a commutative ring and let x be an element of R such that $x^n = 0$ for some positive integer n. Prove that $1 - x$ has a multiplicative inverse in R. [*Hint:* Compute $(1 - x)(1 + x + x^2 + \cdots + x^{n-1})$.]

2. Let x be an element of a ring R such that $x^2 = x$.
   (a) Show that if R is an integral domain, then $x = 0$ or 1.
   (b) Give an example of a ring in which there are elements other than 0 or
   1 for which $x^2 = x$.

3. Let $F = \{0, 2, 4, 6, 8\}$ with addition and multiplication modulo 10.
   (a) Create addition and multiplication tables for F.
   (b) What is the multiplicative identity element in F?
   (c) Argue that F is a field.

4. Let $S = \{(x, y)$, where x and y are integers$\}$ with addition and multiplication
   defined by $(x_1, y_1) + (x_2, y_2) = (x_1 + x_2, y_1 + y_2)$ and $(x_1, y_1) \cdot (x_2, y_2) = (x_1x_2, y_1y_2)$. Show that S is not an integral domain.

5. Let u and v be units in a commutative ring with identity.
   (a) Is $u + v$ a unit? Prove it or give a counterexample.
   (b) Is uv a unit? Prove it or give a counterexample.

6. Let $Z_3[i] = \{a + bi$, where a and b are elements of $Z_3\}$. Clearly, $Z_3[i]$ is a
   ring.
   (a) Is $Z_3[i]$ a field?
   (b) Is it an integral domain?

7. Let $Z_2[x] = \{$polynomials with coefficients in $Z_2\}$.
   (a) Show that every element of $Z_2[x]$ is its own additive inverse.
   (b) Compute the quotient and remainder when $x^5 + x^4 + x^2 + 1$ is divided
   by $x^2 + 1$.
   (c) Prove that if $m = 2^k$, then $(1 + x)^m = 1 + x^m$. [Use induction on k.]
   (d) Is $Z_2[x]$ a field?

8. In Section 7.2, we showed that any element of $Z[\sqrt{2}]$ for which $a^2 - 2b^2 = \pm 1$ is a unit. In this problem, we'll show that these are the *only* units in
   $Z[\sqrt{2}]$.
   (a) Show that u is a unit in $Z[\sqrt{2}]$ iff $x = \dfrac{a}{a^2 - 2b^2}$ and $y = \dfrac{b}{a^2 - 2b^2}$ are
   integers.
   (b) Show that $ax - 2by = 1$.
   (c) Argue that the equation in (b) has a solution in integers only if
   $\gcd(a, b) = 1$.
   (d) If $\gcd(a, b) = 1$, then we know from Lemma 6.1 that there exist integers
   s and t such that $as + bt = 1$. Divide both sides of this equation by $a^2 - 2b^2$
   and conclude that 1 is divisible by $a^2 - 2b^2$.
   (e) Complete the proof.

9. Let R be a ring with multiplicative identity element 1 such that the product of any pair of nonzero elements is nonzero. Show that if a and b are elements of R such that $ab = 1$, then $ba = 1$. [*Hint:* If $ab = 1$, then $aba = a$. Rearrange and factor.]

(*Note:* R is not an integral domain. If it were, then multiplication would be commutative and the result would be trivial.)

10.   When solving a polynomial equation such as $x^2 - 4x + 3 = 0$ in the real numbers, we can factor and set each factor equal to 0. In this case, we get two solutions, $x = 1$ and $x = 3$.

(a) Can we do the same thing if x is an element of a ring? Explain. If no, what other restrictions are necessary?

(b) Solve $x^2 - 4x + 3 = 0$ in Z_7.

(c) Solve $x^2 - 4x + 3 = 0$ in Z_8.

Appendix
An Introduction to Symbolic Logic

A.1. Propositions

By its very nature, a mathematical proof must be logical. In other words, the conclusion must flow from the hypotheses by a series of valid statements. To give you an example of an illogical argument, suppose we wanted to "prove" that it is raining today. Here's the "proof": If it rains, I wear my raincoat. I wore my raincoat today. Therefore it must have been raining.

1. Why is the proof above wrong?

Throughout this text, we have relied on an intuitive sense of logic. While this is probably adequate for most situations, there are occasions where a more formal approach can be useful. That is the purpose of this appendix.

We begin with the following definition.

Definition A *proposition* is a statement that is either true or false.

For example, "George Washington was the first president of the United States," "France is in Asia," and "$37 - 16 = 21$" are all propositions. On the other hand, "$3x + 5 = 13$" is not a proposition, since we don't know the value of x. Other examples of nonpropositions are, "Why is the sky blue?", "Go to the store," and "$25 + 53$."[1]

---

[1] In a different context, the question "Would you like to go on a date?" would be a proposition, but not here.

Propositions can be combined in a variety of ways to form *compound proposi-tions*. Let *p* and *q* be propositions. Then, we have the following definition.

Definition

(a) The *disjunction* of *p* and *q* is the proposition "*p or q*," written $p \lor q$.
(b) The *conjunction* of *p* and *q* is the proposition "*p and q*," written $p \land q$.

For $p \lor q$ to be true, it suffices to have either *p* or *q* or both true. Put differently, the only way in which $p \lor q$ can be false is if both *p* and *q* are false. We can summarize these facts in a *truth table*.

p	*q*	$p \lor q$
T	T	T
T	F	T
F	T	T
F	F	F

In contrast, for $p \land q$ to be true, both *p* and *q* must be true.

2. *Create a truth table for $p \land q$.*
3. *Which of the following propositions are true?*
   (i) "The moon is smaller than the earth and France is in Asia."
   (ii) "The moon is smaller than the earth or France is in Asia."
   (iii) "A bicycle has two wheels or an elm is a type of tree."
   (iv) "A chicken has four legs and $13 > 19$."

We can make more complicated truth tables, involving more than two statements and several operations. For example, here's the truth table for $p \land (p \lor r)$:

p	*q*	*r*	$q \lor r$	$p \land (q \lor r)$
T	T	T	T	T
T	T	F	T	T
T	F	T	T	T
T	F	F	F	F
F	T	T	T	F
F	T	F	T	F
F	F	T	T	F
F	F	F	F	F

We can use truth tables to prove various properties about the disjunction and conjunction operations. For instance, to show that conjunction distributes over disjunction, we would have to show that $p \wedge (q \vee r)$ "is equivalent to" $(p \wedge q) \vee (p \wedge r)$. The truth table for $(p \wedge q) \vee (p \wedge r)$ is:

p	q	r	$p \wedge q$	$p \wedge r$	$(p \wedge q) \vee (p \wedge r)$
T	T	T	T	T	T
T	T	F	T	F	T
T	F	T	F	T	T
T	F	F	F	F	F
F	T	T	F	F	F
F	T	F	F	F	F
F	F	T	F	F	F
F	F	F	F	F	F

Notice that $p \wedge (q \vee r)$ and $(p \wedge q) \vee (p \wedge r)$ have the same truth values for every possible combination of p, q, and r. This means that the two compound propositions are equivalent. Hence, the distributive property holds.

4. *Prove that disjunction distributes over conjunction; that is, show that $p \vee (q \wedge r)$ is equivalent to $(p \vee q) \wedge (p \vee r)$.*

5. *Prove that disjunction is associative; that is, show that $p \vee (q \vee r)$ is equivalent to $(p \vee q) \vee r$.*

Definition   The *negation* of p is the proposition "*not p*," written $\sim p$.

The negation $\sim p$ is true only when p is false. We can form negations either by inserting the word "not" in the appropriate place in the proposition, or by prefacing the proposition with "It is not the case that . . . ." For example, "France is not in Asia," or "It is not the case that giraffes are insects."

If the proposition contains words such as "some" or "all," we have to be careful about where we place the word "not." The negation of "All cars are red" is "Not all cars are red" or "Some cars are not red." It is not "All cars are not red" because that statement and the original proposition are both false. We'll say more about this later when we discuss quantifiers.

Now let's see how negation interacts with disjunction and conjunction. In particular, does negation distribute over disjunction? In other words, is $\sim (p \vee q)$ equivalent to $\sim p \vee \sim q$?

6. Show that the answer to this question is no.

7. Is there a combination of ~ p and ~ q that is equivalent to ~ (p ∨ q)?

8. Is there a combination of ~ p and ~ q that is equivalent to ~ (p ∧ q)?

9. Put your answers to questions 7 and 8 in the following theorem.

Theorem A.1   (DeMorgan's Laws)[2]

(a) ~ (p ∨ q) is equivalent to _____ .
(b) ~ (p ∧ q) is equivalent to _____ .

Theorem A.1 makes sense in words. Let *p* be the proposition "My name is John"; let *q* be the proposition "My sister's name is Helen." Then ~ (p ∨ q) is "It is not the case that my name is John *or* my sister's name is Helen." This can be stated equivalently as "My name is not John *and* my sister's name is not Helen." Similarly, ~ (p ∧ q) is "It is not the case that my name is John *and* my sister's name is Helen." This can be stated equivalently as "Either my name is not John *or* my sister's name is not Helen."

The following two additional ways of forming compound propositions are the underlying principle behind many proofs.

Definition

(a) The *implication* of *p* and *q* is the proposition "If *p*, then *q*." It is written $p \Rightarrow q$.
(b) The *biconditional* (or *double implication*) of *p* and *q* is the proposition "*p* if and only if *q*." It is written $p \Leftrightarrow q$.

In an implication, the proposition *p* is called the *hypothesis*; the proposition *q* is called the *conclusion*. Implication can be expressed in a variety of other ways: "*p* only if *q*," "*p* is sufficient for *q*," or "*q* is necessary for *p*." The truth table for $p \Rightarrow q$ is:

p	q	$p \Rightarrow q$
T	T	T
T	F	F
F	T	T
F	F	T

Notice that the only time $p \Rightarrow q$ is false is when the hypothesis is true and the

---

[2] Augustus DeMorgan (1806–1871) was, along with George Boole (1815–1864), one of the developers of modern symbolic logic. He was born in India, but went to school in England and taught at University College of London.

conclusion is false. Thus, an implication such as "If France is in Asia, then chickens have two legs" is considered to be a true statement because the hypothesis is false and the conclusion is true. On the other hand, "If France is in Europe, then chickens have three legs" is false.

Many mathematical theorems can be stated in the form of implications. One of the first theorems encountered in this book is: "The sum of two even integers is even." This can be restated as: "If x and y are even, then $x + y$ is even." On the other hand, a statement such as, "If $x > 5$, then $x^2 > 25$," is not an implication in the sense in which we have defined implication, because "$x > 5$" and "$x^2 > 25$" are not propositions. (We don't know if they are true or false until we know the specific value of x.)

The biconditional can be expressed as "p is necessary and sufficient for q" or "p is equivalent to q." Its truth table is:

p	q	$p \Leftrightarrow q$
T	T	T
T	F	F
F	T	F
F	F	T

Thus, the biconditional is true only when p and q have the same truth value. We can now express the distributive property proven earlier as $p \wedge (q \vee r) \Leftrightarrow (p \wedge q) \vee (p \wedge r)$.

10. *Make truth tables for each of the following.*
    (i) $p \vee (q \Rightarrow p)$ (ii) $p \Rightarrow (\sim q \vee r)$

Every implication has three other related compound propositions: The *converse* of $p \Rightarrow q$ is $q \Rightarrow p$; the *inverse* is $\sim p \Rightarrow \sim q$; and the *contrapositive* is $\sim q \Rightarrow \sim p$. Observe that the inverse is the contrapositive of the converse.

11. *Prove that an implication and its contrapositive are equivalent; that is, show $(p \Rightarrow q) \Leftrightarrow (\sim q \Rightarrow \sim p)$ is always true.*

12. *Prove that an implication and its converse are not equivalent.*

The result of question 11 allows us to prove theorems by proving their contrapositives.

A compound proposition that is true for every possible truth value of its components is called a *tautology*. The simplest tautology is $p \vee \sim p$, which is true no matter whether p is true or false. An example of this is, "Either it is raining or it is not raining," which is true all the time. Another tautology is $(p \Rightarrow q) \Leftrightarrow (\sim q \Rightarrow \sim p)$.

Tautologies form the basis for methods of proof. Consider the following argument: "If it is Tuesday, I take the train to work. It is Tuesday. Therefore, I take the train to work." If p is the proposition "it is Tuesday" and q is the proposition "I take the train to work," then this argument can be represented symbolically as $[(p \Rightarrow q) \wedge p] \Rightarrow q$.

  13. Prove that $[(p \Rightarrow q) \wedge p] \Rightarrow q$ is a tautology.

  14. Prove that $[(p \Rightarrow q) \wedge \sim q] \Rightarrow \sim p$ is a tautology. Explain this in the context of taking the train to work on Tuesdays.

  15. Prove that $[(p \Rightarrow q) \wedge (q \Rightarrow r)] \Rightarrow (p \Rightarrow r)$ is a tautology. Give an example of this type of argument.

There are many other tautologies we could create.[3]

  The opposite of a tautology—a statement that is always false—is called a *contradiction*.

  16. Give a very simple example (involving just one proposition) of a contradiction.

  17. Is $[(p \Rightarrow q) \wedge q] \Rightarrow p$ a contradiction?

A.2.   Open Sentences and Quantifiers

As we said earlier, an equation such as $3x + 7 = 13$ does not qualify as a proposition because its truth value depends on the unknown quantity x. If $x = 2$, the statement is true; otherwise, it is false.

Definition   A proposition whose truth value depends on one or more variables is called an *open sentence*. The set of values of the variable(s) that make the open sentence true is called the *truth set* of the sentence.

We use the notation $p(x)$ to denote an open sentence whose truth depends on x.

  To determine the truth set of $p(x)$, we must first specify the *universe of discourse*, which is the set of possible values for the variables in the open sentence. For example, let $p(x)$ be the open sentence $x^2 = 2$. If the universe is the set of integers, then the truth set is empty. On the other hand, if the universe is the set of real numbers, the truth set contains two elements, $\sqrt{2}$ and $-\sqrt{2}$.

  1. Graph the truth sets of each of the following open sentences.

---

[3]Some of these tautologies have names. The one in question 13 is called "modus ponens"; the one in question 14 is called "modus tollens"; and the one in question 15 is called "hypothetical syllogism."

(i) (i) $x^2 - 4x - 5 < 0$ *in the universe of real numbers*
(ii) $3x + 2y > 6$ *in the universe of real numbers.*
(iii) $3x + 2y = 6$ *in the universe of integers.*
(iv) $(3x + 2y > 6) \wedge (x - y < 1)$ *in the universe of real numbers.*
(v) $3(x + 2) = 3x + 6$ *in the universe of real numbers.*
(vi) $x^2 < x$ *in the universe of integers.*

Once we assign a value to the variables in an open sentence, we turn it into a proposition that is either true or false. Another way to make a proposition out of an open sentence is to attach a *quantifier* to it.

Definition
(a) The *universal quantifier* $\forall$ indicates that the truth set of an open sentence is the entire universe. The proposition $(\forall x)p(x)$ is read "for all x, $p(x)$."
(b) The *existential quantifier* $\exists$ indicates that the truth set of an open sentence is nonempty. The proposition $(\exists x)p(x)$ is read "there exists an x such that $p(x)$."

Again, we must specify the universe, which we can do by placing it in the quantifier symbol. For example, $(\forall x \in \mathbf{R})p(x)$ means "for all real x, $p(x)$," whereas $(\exists x \in \mathbf{Z})p(x)$ means "there exists an integer x such that $p(x)$."

2. *Which of the following are true?*
(i) $(\exists x \in \mathbf{Z})(4x = 9)$ (ii) $(\exists x \in \mathbf{R})(4x = 9)$
(iii) $(\forall x \in \mathbf{Z})(x^2 \geq x)$ (iv) $(\forall x \in \mathbf{R})(x^2 \geq x)$
(v) $(\exists x, y \in \mathbf{R})(x^2 + y^2 = -1)$
(vi) $(\forall x, y \in \mathbf{R})\left(\dfrac{x^2 - y^2}{x - y} = x + y\right)$

Now let's see how to express mathematical statements in terms of quantifiers. Consider the statement, "All prime numbers are positive integers." We can rewrite this as, "If n is a prime, then n is a positive integer." More precisely, we mean, "For all n (in the universe of real numbers), if n is prime, then n is a positive integer." Symbolically, this can be written as $(\forall n \in \mathbf{R})(n$ is prime $\Rightarrow n$ is a positive integer). Now consider the statement, "Some rational numbers are integers." We can write this as $(\exists x \in \mathbf{Q})(x \in \mathbf{Z})$ or as $(\exists x \in \mathbf{R})(x \in \mathbf{Q} \wedge x \in \mathbf{Z})$.

Here's a more complicated example: "For every pair of distinct rational numbers x and y, there exists a rational number z strictly between x and y." Symbolically, we write $(\forall x, y \in \mathbf{Q})((x \neq y) \Rightarrow (\exists z \in \mathbf{Q})(x < z < y \vee y < z < x))$.

3. *Is this last statement true? Would it be true if we replace "rational number" with "integer"?*

4. *Which of the following are true in the universe of real numbers?*
   (i) $(\forall x)(\exists y)(x + y = 5)$ (ii) $(\forall x)((\exists y)(xy = 5)$
   (iii) $(\exists n \in \mathbf{Z})(\forall x)(x < 0 \Rightarrow x^n > 0)$

5. *Express each of the following in symbolic form.*
   (i)   *Some integers are positive.*
   (ii) *All primes greater than 2 are odd.*
   (iii) *There are no values of x and y such that $x^2 + y^2 = -1$.*

As a final example, consider the definition of $\lim_{x \to a} f(x) = L$. In words, we say "For all $\varepsilon > 0$, there exists $\delta > 0$ such that $|f(x) - L| < \varepsilon$ whenever $0 < |x - a| < \delta$." In symbols, this becomes $(\forall \varepsilon > 0)(\exists \delta > 0)(\forall x)(0 < |x - a| < \delta \Rightarrow |f(x) - L| < \varepsilon)$. Notice that the word *whenever* is another way of expressing the universal quantifier.

Sometimes we need to express the fact that a certain quantity exists and it is the only one with the desired property. In other words, the quantity is *unique*. We use the symbol $\exists!$ to express unique existence. For example, $(\forall x > 0)(\exists! y)(y^2 = x)$ can be translated as "For every positive x, there exists a unique y such that $y^2 = x$."

6. *Is the last statement true? If not, can you fix it so that it is?*

7. *Which of the following are true in the universe of real numbers?*
   (i) $(\exists! x)(2x + 5 = 19)$ (ii) $(\exists! x)(x^2 + 4x = -5)$
   (iii) $(\exists! x)(x^2 + 4x = -4)$ (iv) $(\forall y)(\exists! x)(e^x = y)$

Some of the theorems we have encountered in this book are statements about unique existence. For example, Theorem 1.2 (the Division Algorithm) says "Given any two positive integers a and b, there exist unique nonnegative integers q and r such that $a = bq + r$, where $0 \leq r < b$." If we specify the universe to be $\mathbf{Z}$, then this statement can be written as: $(\forall\, a, b > 0)(\exists! q, r)(q \geq 0 \wedge 0 \leq r < b \wedge a = bq + r)$.

When we prove theorems involving unique existence, we often assume that there are two values x and y that make the proposition true, then show that $x = y$. In symbols, this becomes $(\exists x)(p((x) \wedge (\forall y)[p(y) \Rightarrow x = y])$. Put differently, we show that there exists an x that makes $p(x)$ true and that, whenever $p(y)$ is true, then x must equal y.

Finding the negation of a quantified statement requires some care. Consider the proposition "$p(x)$ is true for all x." To show that this proposition is false, it suffices to produce one x for which $p(x)$ is false. Symbolically, $\sim [(\forall x)p(x)]$ is equivalent to $(\exists x)(\sim p(x))$. So, for example, a negation of "for all x, "$x^2 > x$" is "there is some x such that $x^2 \leq x$." Any value of x that makes $p(x)$ false is called a *counterexample*.

On the other hand, consider the proposition "there exists an x such that $p(x)$ is true." To show that this is false, we have to show that $p(x)$ is false for every x. Symbolically, $\sim [(\exists x)p(x)]$ is equivalent to $(\forall x)(\sim p(x))$. For example, a negation of "there exists x and y such that $x^2 + y^2 = -1$" is "for all x and y, $x^2 + y^2 \neq -1$."

Finally, to negate a statement of unique existence, we have to show either that $p(x)$ is false for all x or that there exists $y \neq x$ such that $p(y)$ is true. That is, $(\forall x)(\sim p(x) \vee (\exists y)[p(y) \wedge x \neq y])$.

A.3. Additional Questions

1. The sign, "No Shirt, No Shoes, No Service," often appears in restaurants. There are many possible interpretations to this sign, such as "shirt, shoes, and service not sold" or "shirt, shoes, and service not permitted." Give a reasonable interpretation of this sign and express it symbolically.

2. Determine a negation of $p \Rightarrow q$ and express it in terms of disjunction and/or conjunction.

3. We defined the disjunction of p and q to be true whenever p or q or both are true. Another type of disjunction, often called *exclusive disjunction,* is true when p or q but *not both* are true. Let's denote this operation by $p * q$.

(a) Express $p * q$ in terms of the usual conjunction and disjunction and make a truth table for it.

(b) Is $*$ associative? Prove or disprove.

4. The Principle of Mathematical Induction says that if an open sentence is true for $n = 1$ and if it is true for $n = k + 1$ whenever it is true for $n = k$, then it is true for all positive integers n. Express this principle symbolically.

Index

A

Abundant numbers, 27
Additive identity, 142
Additive inverse, 142
Algebraic numbers, 70
Algorithm, 5
Antidifference of a sequence, 123
Argument of a complex number, 58
Arithmetic-Geometric-Mean Inequality, 116
Arithmetic mean, 115
Arithmetic sequence, 11, 12
Associate of an element of a ring, 148
Associative property, 2, 142

B

Base of a numeration system, 44
Biconditional, 164
Binet's formula for Fibonacci numbers, 100
Binomial coefficients, 76
Binomial theorem, 78

C

Canonical representation of an integer, 21
Chaos, 108
Characteristic equation, 99
Chinese Remainder Theorem, 130
Closure property, 1, 142
Cobweb diagram, 104
Combinations, 79
Combinatorics, 75
Commutative property, 2, 142

Complex numbers, 58
  Arithmetic operations, 59
  Polar representation, 58
Complex integers, 151
Conclusion, 3, 164
Congruences, 6, 7, 127
Conjugate of a complex number, 60
Conjunction, 162
Constructible numbers, 68, 71
Content of a polynomial, 58
Continued fractions, 37
  Finite, 37
  Infinite, 37, 41
  Periodic, 42
  Representation of rational numbers, 38
  Simple, 37
Continued radical, 48
Contradiction, 37, 165
Contrapositive of a statement, 20, 162
Convergence of a series, 13
Convergents of a continued fraction, 43
Converse of a statement, 20, 165
Counterexample, 4, 168
Cycle, 109
Cyclic behavior, 106, 109

D

de la Vallee Poussin, Charles, 19
Decimal representation, 31
  Repeating, 31
  Terminating, 31
Deficient numbers, 27

Degree
  of an algebraic number, 70
  of a polynomial, 52
  of a vertex in a graph, 90
DeMoivre's Theorem, 60
DeMorgan, Augustus, 164
  DeMorgan's Laws, 164
Descartes' Rules of Signs, 63, 64
Difference equation, 97
  General solution, 98
  Linear, homogeneous, 97, 98
  Particular solution, 98
Difference of a sequence, 122
Diophantine equation, 125
  and congruences, 127
Dirichlet, Peter Gustav, 133
Disjunction, 162
Distributive property, 2, 142
Divisibility, 9
Division Algorithm, 4
  for integers, 5
  for polynomials, 55
Divisors of an integer, 22
  Number of, 22
  Sum of, 23
Divisor of zero, 145
Duplicating the cube, 68

E

Edge of a graph, 88
Eisenstein Irreducibility Criterion, 87
Elliptic curves, 133
Euclid, 18
Euclidean domain, 150
Euler, Leonhard, 89
Euler cycle, 89
Euler-Fermat Theorem, 33, 34
Euler path, 89
Euler's formula, 92
Euler's phi function (totient function), 33
Existence and uniqueness theorem, 5
Existential quantifier, 167

F

Factoring polynomials, 55
Factor theorem, 56
Fermat's Last Theorem, 133
Fermat's Little Theorem, 134, 135
Fibonacci (Leonardo of Pisa), 14

Fibonacci sequence, 14, 100
Field, 155
Fixed point, 103
  Attracting, 105
  Neutral, 106
  Repelling, 105
Fractional part of a number, 120
Function iteration, 101
  Linear, 101
  Nonlinear, 106
Fundamental Theorem of Arithmetic, 20, 149

G

Gauss, Karl, 12
Gaussian integers, 151
Generating function, 84, 85
Geometric constructions, 67
Geometric mean, 115
Geometric sequence, 11, 12
Graph, 88
  Planar, 91
Graph theory, 75
Greatest common divisor, 25
Greatest integer function, 119

H

Hadamard, Jacques, 19
Harmonic mean, 117
Hermite, Charles, 66
Hypothesis, 3, 164
Hypothetical syllogism, 166

I

Implication, 164
Induction hypothesis, 16
Integers, 1
  Even and odd, 2
Integral domain, 145, 155
Inverse of a statement, 20, 165
Irrational numbers, 30, 35
Irreducible ,
  element of a ring, 149
  polynomial, 55

K

Königsburg Bridge Problem, 87
Kummer, Ernst, 133, 156
Kuratowski's Theorem, 91

L

Lame, Gabriel, 133
Least common multiple, 25
Legendre, Adrien-Marie, 133
Lehrer, Tom, 46
Lindemann, Ferdinand, 66
Liouville, Joseph, 66
Lucas, Edouard, 101
   Lucas sequence, 101
   Lucas-Lehmer test, 101

M

Magnitude of a complex number, 58
Mathematical induction, principle of, 15
Mersenne, Father Marin, 25
   Mersenne primes, 25
Modulo (modulus), 7
Modus ponens, 166
Modus tollens, 166
Multiplicative identity, 142
Multiplicative inverse, 143

N

Natural numbers, 155
Negation, 163
Numeration systems, 43
   Conversion of integers, 45
   Conversion of rational numbers, 46
   Positional, 44

O

Open sentence, 166

P

Pascal, Blaise, 77
   Pascal's triangle, 77
Perfect numbers, 24
Polyhedron, 92
   Regular, 92
Polynomial (function), 51
   Arithmetic operations, 53
   Reciprocal, 72
   Zero polynomial, 53,
Prime Number Theorem, 19
Prime numbers, 17, 141
Proposition, 161
   Compound, 162

Pythagoras, 132
   Pythagorean Theorem, 130
   Pythagorean triplet, 130
     Primitive, 130

Q

Quadratic formula, 62
Quadratic surd, 42

R

Rational numbers, 29
   Decimal representation, 31
   Extensions, 36
Rational roots theorem, 65
Ring, 142
   Commutative, 142
   With identity, 142
Root-mean-square, 118
Roots of polynomial equation, 56, 61

S

Sequences, 10
   Explicit formula, 10
   Recurvsive formula, 10
Series, 11
Sieve of Eratosthenes, 17, 18
Sigma notation, 11
Squaring the circle, 68
Summation by parts, 125

T

Tautology, 166
Taylor series, 86, 87
Transcendental number, 66
Trisecting an angle, 68
Truth table, 162

U

Unique existence, 169
Unique factorization, 22, 141, 147
   for integers, 22
   for polynomials, 56
Unique factorization domain, 149
Unit of a ring, 147
Unit's digits, 134
Universal quantifier, 167
Universe of discourse, 166

V

Vertex of a graph, 88

W

Well-Ordering Principle, 5, 27
Wiles, Andrew, 133

Z

Zeroes of a polynomial, 56, 61
$Z_{m, 8, 128}$